aus der Reihe:

Innovationen mit Mikrowellen und Licht

**Forschungsberichte aus dem Ferdinand-Braun-Institut,
Leibniz-Institut für Höchstfrequenztechnik**

Band 55

Simon Rauch

Lateral emission characteristics of high-power broad-area lasers
subject to external optical feedback

Herausgeber: Prof. Dr. Günther Tränkle, Prof. Dr.-Ing. Wolfgang Heinrich

Ferdinand-Braun-Institut   Tel.   +49.30.6392-2600
Leibniz-Institut   Fax   +49.30.6392-2602
für Höchstfrequenztechnik (FBH)
Gustav-Kirchhoff-Straße 4   E-Mail fbh@fbh-berlin.de
12489 Berlin   Web   www.fbh-berlin.de

Innovations with Microwaves and Light

Research Reports from the Ferdinand-Braun-Institut, Leibniz-Institut für Höchstfrequenztechnik

Preface of the Editors

Research-based ideas, developments, and concepts are the basis of scientific progress and competitiveness, expanding human knowledge and being expressed technologically as inventions. The resulting innovative products and services eventually find their way into public life.

Accordingly, the *"Research Reports from the Ferdinand-Braun-Institut, Leibniz-Institut für Höchstfrequenztechnik"* series compile the institute's latest research and developments. We would like to make our results broadly accessible and to stimulate further discussions, not least to enable as many of our developments as possible to enhance everyday life.

High-power broad-area laser diodes are key light sources for high-brightness lasers used in materials processing. The performance of such laser systems can be significantly enhanced when using broad-area lasers that are spectrally stabilized by external optical feedback. So far, however, these coupled-cavity systems suffer from an increased degradation of lateral beam quality and lifetime. This thesis demonstrates the strong influence of the spatial distribution of the internal temperature on these parameters. The investigations have enabled a tailored device design with a more homogenous longitudinal intensity profile, leading to improved lateral emission characteristics.

We wish you an informative and inspiring reading

Prof. Dr. Günther Tränkle
Director

Prof. Dr.-Ing. Wolfgang Heinrich
Deputy Director

The Ferdinand-Braun-Institut

The Ferdinand-Braun-Institut researches electronic and optical components, modules and systems based on compound semiconductors. These devices are key enablers that address the needs of today's society in fields like communications, energy, health and mobility. Specifically, FBH develops light sources from the visible to the ultra-violet spectral range: high-power diode lasers with excellent beam quality, UV light sources and hybrid laser systems. Applications range from medical technology, high-precision metrology and sensors to optical communications in space and integrated quantum technology. In the field of microwaves, FBH develops high-efficiency multi-functional power amplifiers and millimeter wave frontends targeting energy-efficient mobile communications as well as car safety systems. In addition, compact atmospheric microwave plasma sources that operate with economic low-voltage drivers are fabricated for use in a variety of applications, such as the treatment of skin diseases.

The FBH is a competence center for III-V compound semiconductors and has a strong international reputation. FBH competence covers the full range of capabilities, from design to fabrication to device characterization.

In close cooperation with industry, its research results lead to cutting-edge products. The institute also successfully turns innovative product ideas into spin-off companies. Thus, working in strategic partnerships with industry, FBH assures Germany's technological excellence in microwave and optoelectronic research.

Lateral emission characteristics of high-power broad-area lasers subject to external optical feedback

vorgelegt von
M.Sc.
Simon Rauch
geb. in Marburg

an der Fakultät IV – Elektrotechnik und Informatik
der Technischen Universität Berlin
zur Erlangung des akademischen Grades

Doktor der Naturwissenschaften
– Dr. rer. nat. –

genehmigte Dissertation

Promotionsausschuss:

Vorsitzender:	Prof. Dr. Wolfgang Heinrich
Erstgutachter:	Prof. Dr. Günther Tränkle
Zweitgutachter:	Prof. Dr. Thomas Dekorsy
Drittgutachter:	Prof. Dr. Michael Kneissl

Tag der wissenschaftlichen Aussprache: 13. Dezember 2019

Berlin, 2020

Bibliografische Information der Deutschen Nationalbibliothek

Die Deutsche Nationalbibliothek verzeichnet diese Publikation in der Deutschen Nationalbibliographie; detaillierte bibliographische Daten sind im Internet über http://dnb.d-nb.de abrufbar.

1. Aufl. - Göttingen: Cuvillier, 2020

    Zugl.: Berlin, Technische Universität, Diss., 2019

© CUVILLIER VERLAG, Göttingen 2020

Nonnenstieg 8, 37075 Göttingen

Telefon: 0551-54724-0

Telefax: 0551-54724-21

www.cuvillier.de

1. Auflage 2020

Gedruckt auf umweltfreundlichem, säurefreiem Papier aus nachhaltiger Forstwirtschaft.

ISBN 978-3-7369-7158-5

eISBN 978-3-7369-6158-6

Abstract

The beam quality and lifetime of broad-area laser diodes degrades with increasing optical output power. In particular, towards high optical output powers, the near field of gain-guided broad-area lasers is progressively narrowed by thermo-optic effects which leads to an enhanced optical facet load and an over-proportionate far-field blooming. So far, counter-measures have focused on various modifications of the lateral device structure to tailor the lateral gain, temperature and intensity profiles which all are accompanied by a decrease of optical power and electro-optical conversion efficiency.

Common single-side emitting broad-area lasers are extremely asymmetric devices with respect to the longitudinal direction: The high semiconductor gain along with a high single-side output coupling leads to a strong asymmetry of the propagating fields. Hence the internal optical power and thus the thermal load are highest in the vicinity of the out-coupling facet for free-running operation. Moreover, external optical feedback, which is frequently applied intentionally for spectral stabilization, but may also be present by unintentional back-reflections, intensifies the optical and thermal load at the emission facet of the diode laser and might even change the power asymmetry.

In this thesis, the impact of the longitudinal distribution of internal intensity, temperature and gain on the lateral beam parameters is revealed. Via numerical simulations it is demonstrated that homogenizing the longitudinal device temperature prevents near-field narrowing and accompanying far-field blooming which is experimentally verified by symmetrizing the facet reflectivities leading to double-side emitting devices. The accompanying homogenization of the longitudinal intensity profile has enabled a simultaneous improvement of both the electro-optical efficiency and the lateral beam parameter product in the order of $1\,\text{mm} \times \text{mrad}$ for an emission stripe width of $140\,\mu\text{m}$.

External-cavity broad-area lasers show a significantly higher degradation of laser lifetime and beam quality compared to free-running devices, which cannot solely be explained by the additional light power re-injected into the waveguide. Therefore, the second part of this thesis deals with the impact of spatially mismatched optical feedback on beam quality and lifetime.

For the investigated InGaAs/AlGaAs-based, gain-guided structure, it is found that irradiance of solder and p-side metalization as well as the adjacent highly p-doped semiconductor layers causes a localized heating around the front facet which leads to a strong reduction of the threshold for catastrophic optical damage. Furthermore, it is revealed that lateral displacement of the feedback spot can lead to a beam quality degradation for single-side emitting devices of over $40\,\%$, which could be reduced by symmetrically out-coupling external-cavity configurations.

Kurzfassung

Die Strahlqualität und Lebensdauer von Breitstreifendiodenlasern degradiert mit zunehmender optischer Ausgangsleistung. Besonders das Nahfeld von gewinngeführten Breitstreifenlasern zieht sich für hohe Ausgangsleistungen zunehmend durch thermo-optische Effekte zusammen, was zu einer erhöhten Facettenlast und überproportionaler Fernfeldverbreiterung führt. Bisherige Gegenmaßnahmen bezogen sich auf verschiedenste Modifikationen der lateralen Bauteilstruktur, um die laterale Verteilung des Gewinns, der Temperatur und der Intensität anzupassen, welche mit Verlusten in der optischen Leistung und elektro-optischen Konversionseffizienz einhergehen.

Gewöhnliche einseitig-emittierende Breitstreifenlaser sind Bauteile, die sehr asymmetrisch in der longitudinalen Richtung sind: Der hohe Gewinn des Halbleitermediums gepaart mit einem hohem einseitigen Auskoppelgrad führt zu einer starken Asymmetrie der propagierenden Felder. Daher sind im freilaufenden Betrieb die optische Leistung und thermische Last am höchsten in der Umgebung der Auskoppelfacette. Darüber hinaus erhöht optische Rückkopplung, welche häufig zur spektralen Stabilisierung verwendet wird aber auch durch ungewollte Rückreflexe auftreten kann, die thermische und optische Last an dieser Facette und kann auch die Leistungsasymmetrie beeinflussen.

In dieser Arbeit wird der Einfluss der longitudinalen Verteilung von interner Intensität, Temperatur und Gewinn auf die lateralen Strahlparameter aufgezeigt. Es wird mithilfe numerischer Simulationen gezeigt, dass eine Homogenisierung der longitudinal Bauteiltemperatur eine Nahfeldverengung und damit verbundene Fernfeldverbreiterung unterbinden kann, was experimentell anhand von doppelseitig auskoppelnden Bauteilen, erreicht durch Angleichen der Facettenreflektivitäten, demonstriert wird. Durch die damit einhergehende Homogenisierung der longitudinalen Intensitätsverteilung konnte eine simultane Verbesserung der elektro-optischen Effizienz und des lateralen Strahlparameterprodukts im Bereich von $1\,\mathrm{mm} \times \mathrm{mrad}$ für eine Emissionsstreifenbreite von $140\,\mu\mathrm{m}$ erreicht werden.

Breitstreifenlaser mit externem Resonator zeigen eine signifikant höhere Degradation von Lebensdauer und Strahlqualität als im freilaufenden Betrieb, was nicht alleinig auf die zusätzlich in den Wellenleiter zurückgekoppelte optische Leistung erklärt werden kann. Deshalb behandelt der zweite Teil dieser Arbeit den Einfluss von räumlich fehlangepasster optischer Rückkopplung auf Strahlqualität und Lebensdauer.

Für die untersuchte gewinngeführte Struktur basierend auf InGaAs/AlGaAs wurde herausgefunden, dass eine Bestrahlung des Lots zusammen mit der Metallisierung der p-Seite und den angrenzenden hochdotierten Halbleiterschichten eine Erwärmung der Frontfacette hervorruft, was zu einer starken Reduktion der Schwelle für die abrupte optische Zerstörung des Bauteils führt. Darüber hinaus wird demonstriert, dass ein lateraler Versatz des zurück auf die

Laserdiode reflektierten Lichts zur einer Strahlqualitätsverschlechterung von über 40 % für einseitig auskoppelnde Emitter führen kann, was durch symmetrische Auskopplung reduziert werden konnte.

Contents

1. Introduction

High-power broad-area laser diodes (BALs) are the most efficient devices to convert electrical into optical power [1]. Maximum peak efficiencies for free-running continuous-wave (CW) operation at room temperature reach over 70 % [2–5]. These extraordinary efficiency properties can be obtained with almost no penalty for wavelength stabilization of BALs by an external cavity (external-cavity BALs (EC-BALs)) that provides narrow-band optical feedback [6]. Such favorable characteristics along with their compact size and high integrability make BALs key components of efficient high-brightness solid-state lasers for materials processing. They are indispensable as pump sources for Yb-based fiber and disk lasers and as beam sources in high-brightness direct diode lasers. Prominent applications for such lasers are flat-sheet metal cutting and remote welding.

The widespread use of diode laser pump sources is due to their narrow-band, high-power radiation which has enabled room-temperature pumping of three-level solid-state lasers [7] that exhibit a low Stokes shift and hence low thermo-optical aberrations. Thereby inefficient lamp-pumped lasers have successively been replaced by diode-pumped solid state lasers (DPSSLs) during the last decade.

In 2017 Yb-doped DPSSLs had a 60 % market share of industrial lasers revenues [8]. This high position has evolved due to the fact that the broadband absorption band of several Yb-doped solids from around 900 nm to 950 nm overlaps with the emission spectral range of reliable high-power InGaAs/AlGaAs-based semiconductor lasers. As a consequence, Yb-doped gain media can be pumped by free-running BALs which allows compact and robust pump source architectures. Moreover, the close proximity of this broadband absorption band of Yb to the corresponding most widely used lasing transitions around 1030 nm and 1070 nm leads to a low quantum defect of less than 15 % and thus to a high optical-to-optical conversion efficiency.

Despite its high degree of sophistication, the technology of Yb-doped DPSSLs can be further optimized concerning efficiency with wavelength-stabilized EC-BALs. The latter may be used to pump the narrow-band zero-phonon transition of Yb-doped gain media close to a wavelength of 970 nm, which effectively reduces the remaining quantum defect to $\sim$6 %, when lasing around 1030 nm is assumed.

Disk lasers benefit especially from a reduction of the quantum defect by zero-phonon line pumping. Waste heat generated by the Stokes shift is the dominant cause of beam quality deterioration, preventing single-mode output at multi-kW power levels [9]. Moreover, for Yb-doped fiber lasers, the pump light absorption is maximum for the zero-phonon line so that pumping at this wavelength helps to reduce the fiber length and hence to shift the onset of non-linearities like stimulated Brillouin and Raman scattering towards higher output powers [10].

By definition, DPSSLs exhibit a lossy conversion of pump to lasing photon. From a physical point of view, a direct diode laser that omits such a conversion process and uses the BAL light directly for materials processing, has the potential to become an even more efficient laser platform. Throughout the last 10 years, direct diode lasers, that are based on dense wavelength beam combining (DWBC) of hundreds of wavelength-stabilized EC-BALs into a single high-brightness output beam, have become feasible. The most promising technique in terms of brightness and robustness employs a single external grating-cavity that is simultaneously used for wavelength stabilization and beam combination [11, 12]. Most recent results demonstrate an electrical-to-optical (e-o) efficiency of 50 % at an output power of 2 kW [13], which is already close to the level of the most efficient fiber lasers [14], but still about 10 percentage points less than the efficiency of free-running BAL bars. This difference indicates that state-of-the-art direct diode lasers have the potential for further enhancement of e-o efficiency.

Despite the obvious physical advantages of employing EC-BALs for efficient high-power lasers, the industrial use of EC-BALs is problematic. The main technological challenge is that optical feedback accelerates long-term device degradation [15] and thus reduces the onset time for catastrophic optical damage (COD) [16]. As yet, these technological challenges outweigh the physical advantage of an improved efficiency.

Nearly all industrial high-power diode lasers are based on GaAs quantum well (QW) technology embedded in an index waveguide. This combination provides low optical and carrier losses along with nearly diffraction-limited beam quality in the vertical direction [17]. In order to meet industry lifetime requirements of several 10 kh, the design of high-power semiconductor lasers aims at keeping the optical intensity and the change of internal temperature as low as possible. Commonly, large longitudinal cavity lengths (3 to 6 mm) are employed to reduce thermal and electrical resistances. In order to balance the optical load, large vertical cavity designs with effective mode diameters in the order of 1 µm [18] or larger are used. Additionally, the output facets require supplementary passivation techniques since they are exposed to the highest intensities inside the laser and are susceptible to surface states of enhanced non-radiative recombination and absorption.

An additional characteristic of BALs is a wide lateral aperture (typically several 10 µm to several 100 µm), as illustrated in Fig. 1.1. The expansion of

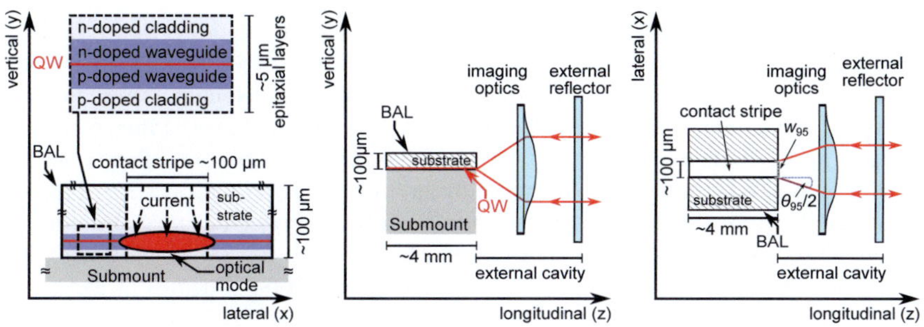

Fig. 1.1.: Profile views of a typical EC-BAL.

the lateral waveguide results in a further decrease of the thermal and electrical resistances and relaxes the optical load at the out-coupling mirror compared to fundamental-mode ridge-waveguide lasers with a typical lateral aperture size of $<10\,\mu$m. Because of that, the peak efficiency of state-of-the-art ridge-waveguide lasers is about 10 percentage points less than for BALs [19]. BALs manage to deliver output powers of 10 W and intensities of $\sim$10 MW cm^{-2} reliably with an estimated mean time to failure of about 25 years [20]. However, the concept of broadening the lateral waveguide is always a trade-off between power and efficiency on one side and beam quality on the other side. For the wider the lateral waveguide, the more higher-order transverse-lateral modes are supported, degrading the beam quality as quantified by an increasing lateral beam parameter product (BPP$_{\mathrm{lat}}$) [21, 22]. Here, BPP$_{\mathrm{lat}}$ is proportional to the product of lateral near- and far-field widths. Consequently, the lateral stripe width of BALs is chosen according to the required BPP$_{\mathrm{lat}}$ by a specific application. Furthermore, in the lateral direction, built-in index trenches close to the contact stripe lead to a far-field increase compared to gain-guided BALs [23]. This implies that, for a fixed aperture width, the lateral emission characteristics of gain-guided BALs are governed by thermal and carrier-induced refractive index changes.

Alternative approaches such as monolithic master-oscillator-power-amplifier configurations try to both circumvent multi-mode emission and relax the facet load by increasing and amplifying the fundamental mode from a ridge laser via a tapered amplifier section [24]. Based on such technology, output powers >5 W with nearly diffraction-limited beam quality have been demonstrated [25]. However, due to a comparatively large taper angle in the range of 6°, these devices exhibit an astigmatism that depends strongly on the operation current [26].

To utilize the advantages of a rectangular broad-stripe waveguide, but simultaneously avoid the mechanical complexity associated with an external cavity design, monolithic alternatives to EC-BALs have been proposed. The most promising alternative approach is the distributed feedback (DFB)-BAL that employs intra-cavity etched Bragg gratings for wavelength stabilization. However, the etching process is rather complex and tends to introduce contamination and scattering losses [27]. As a consequence, the peak efficiencies of DFB-BALs are typically 5 to 10 percentage points lower compared to Fabry-Pérot BALs [28]. Moreover, the thermally induced spectral emission shift is about 50 times larger for DFB-BALs [29] compared to EC-BALs stabilized by thin-film filters [30] or diffraction gratings [11] and about 5 times larger than for EC-BALs stabilized by a volume Bragg grating (VBG) [31]. This is why only BALs that are stabilized externally are competitive for direct diode lasers based on DWBC.

The commercial success of BALs began with the application of double-clad fibers in 1988 [32] allowing efficient pumping of rare-earth doped fiber lasers and amplifiers with high-power multi-mode beams. In the beginning, primarily AlGaAs-based BALs in the 800-nm range were used for pumping Nd-based lasers and, later, InGaAs-based lasers in the 9xx-nm range for Yb. This transition resulted not only in a lower quantum defect, but also in a boost for reliability. It has been demonstrated that including indium in the quantum well increases the laser lifetime both in free-running [33] and external-feedback operation [34] by almost one order of magnitude.

At the end of the 1990s, the maximum CW power density before catastrophic optical mirror damage (COMD) was in the range of $18\,\mathrm{MW\,cm}^{-2}$ from a $100\,\mu\mathrm{m}$-wide stripe. Steady improvement of facet protection technology [35–37] has lead to the situation that the maximum reported CW power amounts to $23.1\,\mathrm{MW\,cm}^{-2}$ [38] which was limited by thermal rollover instead of COMD.

In the 1990s, research on BALs mainly focused on understanding and mitigation of their multi-modal behavior and filamentation [39, 40] and accompanying non-thermal spatio-temporal instabilities [41–44]. The research was driven by the desire to restrict the high-power, multi-modal beam of BALs to fundamental-mode emission and simultaneously avoid or at least stabilize filamentation. In the early 2000s, external-cavity configurations using lateral fourier-space filtering seemed promising candidates to suppress higher-order transversal modes [45–48]. However, these approaches are always accompanied by a high efficiency loss and restricted to a narrow interval of operation – mostly close to threshold. This is due to the high semiconductor gain combined with self-heating induced waveguiding, commonly referred to as thermal lensing, helping high-order lateral modes to reach threshold.

In the following years, theoretical investigations of high-power semiconductor lasers [49–53] incorporated both thermally and carrier-induced refractive index changes and their respective impact on beam quality to investigate beam quality deterioration with increasing output power [21, 54]. All applications have a certain limit of acceptable BPP_{lat}, which limits usable power extraction from BALs. However, limited quantitative agreement with experimental results made it difficult to identify the root causes of beam quality degradation unambiguously and quantitatively – which prompted additional experimental studies [23, 55].

In parallel but mostly independent of the investigations regarding beam quality degradation, extensive research on the dynamics of COD in laser diodes [56–61] was conducted, which, in case of sufficient cooling, represents the second main limitation of scaling the output power beside lateral beam quality deterioration. These studies on COD emphasize the importance of the temperature dependence of loss and thermal conductivities in order to reach the melting temperature of the active material and thus COD. Due to a detailed consideration of these computation-intensive thermal effects, most papers do not resolve the inputs for the thermal solver, carrier and photon distributions, in both lateral and longitudinal direction.

The past studies have revealed that increased BAL self-heating is a driving force for both beam quality deterioration and COD. However, for state-of-the-art InGaAs-based BALs in the 9xx-nm range, there is little known about the contribution of lateral-longitudinal variations of the internal temperature distribution to these two limitations. Especially, the analysis of EC-BALs requires exactly such a spatial resolution: The high load on the front facet is increased by light coupled back to the diode and potential mismatched parts of the feedback light strongly influence the temperature distribution locally [62].

The present thesis deals with the analysis and tailoring of the lateral beam parameters of EC-BALs as a function of the lateral-longitudinal internal temperature distribution. The analysis focuses on quasi gain-guided InGaAs-based BALs with a lateral stripe width of 100 to 140 µm that are usually used for fiber laser pumping and direct diode lasers.

In chapter 2, free-running operation and laser operation with matched optical feedback is considered. After a brief introduction of the employed experimental setup, laser structure and simulation tools, an analysis of the longitudinal temperature distribution follows: A simplified 2D lateral-vertical temperature solver is coupled slice-wise to the lateral-longitudinal carrier and photon profiles to estimate thermal lensing and its impact on near and far-field widths at both output facets (cf. sect. 2.4) of standard single-side outcoupling BALs. In section 2.5, two-side out-coupling by means of equal facet reflectivities is introduced as a means to tailor the lateral beam parameters, the temperature profile as well as lateral carrier accumulation. With this approach, the experimental feasibility of longitudinal intensity symmetrization by external optical feedback is analyzed subsequently.

In chapter 3, spatially mismatched feedback is investigated. The impact of absorption of feedback light by the adjacent layers surrounding the waveguide on facet heating and COD is studied in section 3.1. In the subsequent section, lateral displacement of the feedback return spot is discussed with respect to beam quality degradation.

The concluding chapter 4 provides an overview of the achievements of this thesis and a brief outlook for future investigations.

Parts of the present thesis have been published:

Journals

- S. Rauch, H. Wenzel, M. Radziunas, M. Haas, G. Tränkle, and H. Zimer, "Impact of longitudinal refractive index change on the near-field width of high-power broad-area diode lasers," *Applied Physics Letters*, vol. 110, no. 26, p. 263504, 2017.

- S. Rauch, C. Holly, and H. Zimer, "Catastrophic optical damage in 950-nm broad-area laser diodes due to misaligned optical feedback and injection" *IEEE Journal of Quantum Electronics*, vol. 54, pp. 1–7, Aug 2018.

Conference Proceedings

- S. Rauch, M. Haas, and H. Zimer, "Numerical and experimental investigation of near-field narrowing in broad-area laser diodes due to longitudinally asymmetric self-heating" in *2017 IEEE Photonics Conference (IPC)*, pp. 111–112, Oct 2017.

- S. Rauch, P. Modak, C. Holly, and H. Zimer, "Beam quality improvement of broad-area laser diodes by symmetric facet reflectivities," in *2018 IEEE International Semiconductor Laser Conference (ISLC)*, pp. 1–2, Sept 2018.

2. BALs subject to matched optical feedback

In this chapter the steady-state lateral-longitudinal temperature and intensity distributions of EC-BALs and their influence on the lateral beam parameters and lifetime is evaluated. Subsequently, a tailoring of the internal temperature and intensity profiles is conducted via facet reflectivity variation.

Regarding laser lifetime and reliability, the catastrophic optical (mirror) damage (CO(M)D) in the active zone of the semiconductor is the dominant thermally activated failure mode. Its temporal evolution can be subdivided into three phases, following the categorization by Hempel [60]: In the first phase, the potential damage volume is heated till it reaches a certain critical temperature (around $150\,^\circ$C). Here, the time until the critical temperature is reached can vary from some nanoseconds during short-pulse current injection, to several years under CW operation. Once the critical temperature is reached, the second phase, the so-called thermal runaway sets in, in which the material heats up rapidly (on nanosecond timescale) beyond its melting temperature. In the melted area the epitaxial layer structure is destroyed and thus defects are created that spread during the third phase referred to as dark-line defects (DLDs).

Due to the ongoing industrial demand for high-power diode lasers with improved beam quality, the CO(M)D remains still a critical issue despite state-of-the-art facet passivation technology: Continuous defect accumulation during long-term degradation eventually ends in CO(M)D after a time period $t_{\mathrm{CO(M)D}}$ when the usable optical output power suddenly drops close to zero. Because of its thermal nature, $t_{\mathrm{CO(M)D}}$ can be expressed in terms of an Arrhenius law where the temperature can be considered proportional to the power per stripe width [63].

In fact, for each output facet, the lateral power confinement is quantified by the near-field width w_{95} which, along with the corresponding far-field width θ_{95}, determines $\mathrm{BPP_{lat}}$ by [23]

$$\mathrm{BPP_{lat}} = w_{95}\theta_{95}/4. \tag{2.1}$$

Here, the lateral beam parameters w_{95} and θ_{95} include $95\,\%$ power content (p.c.). This specific choice of power inclusion is arbitrary, but well-established in diode-laser industry.

Equation (2.1) indicates the dilemma between reliability and beam quality: While a larger near-field width is beneficial for laser lifetime, it increases

BPP_{lat} and thus reduces beam quality. Hence a compromise between BPP_{lat} and $t_{CO(M)D}$ has to be found for each application of BALs individually. In first approximation, w_{95} is determined and hence can be tailored by the contact stripe width w_c.

For gain-guided BALs at high-power operation, however, w_{95} can be significantly lower than w_c, which is referred to as near-field narrowing.

There is experimental indication [53, 64] that near-field narrowing is connected to a strong thermally induced lateral waveguide, often referred to as thermal lensing: As illustrated in Fig. 1.1, BALs are soldered with the epitaxial side down to the submount which is attached to a heat sink (not shown). Additionally, the lateral width of the contact stripe amounts to only about 2 % of the lateral submount width and to about 25 % of the vertical submount height. Hence, the waste heat, which is mainly generated in the active and waveguide layers within the area of the contact opening, spreads both in vertical and lateral direction towards the heat sink. As a major consequence of the lateral component of the heat flow, a thermally induced refractive index profile similar to a bell-shape builds up in the lateral direction.

So far, reported comparisons of experimental data of the lateral beam parameters with numerical simulations have either neglected longitudinal variations [65] or have restricted the benchmark to front-facet data only [53]. Yet in common single-side emitting BALs, the internal intensity distribution and hence the heat power density due to optical absorption are highly asymmetric with regard to the longitudinal direction [66]: Thermal lensing increases along the resonator towards the output facet. In particular, optical feedback changes the internal intensity by power re-injection and excitement of a different lateral mode set compared to the free-running operation [67]. These facts give a clear indication that the longitudinal temperature profile impacts the beam parameters θ_{95} and especially w_{95} differently for each facet and thus has to be considered for the estimation of $t_{CO(M)D}$.

Therefore, in this chapter the influence of the longitudinal temperature distribution on the beam parameters is investigated by means of numerical simulations in comparison with experimental results. Afterwards, w_{95} and hence BPP_{lat} are tailored by means of longitudinal intensity and temperature symmetrization.

2.1. Model of matched optical feedback

This section briefly introduces an idealization of optical feedback where perfect spatial and spectral back-coupling is assumed.

Figure 2.1 depicts the cavity model used for the analysis of BALs subject to optical feedback throughout this chapter. The laser system is modeled

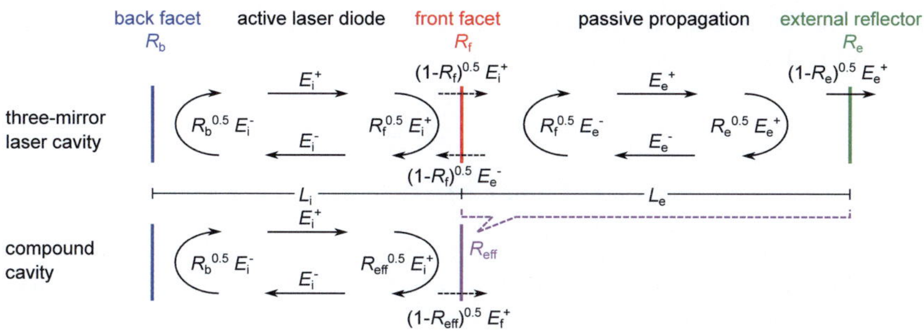

Fig. 2.1.: Three-mirror model for a Fabry-Pérot diode laser coupled to an external cavity via the front facet (top) and equivalent two-mirror system with an effective front-facet reflectivity R_{eff} (bottom).

via a simplified three-mirror system consisting of an internal active resonator of length L_i and an external passive cavity of length L_e. The forward and backward traveling electric fields of the internal resonator $E_i^{\pm}$ and the external cavity $E_e^{\pm}$ are coupled via the anti-reflection (AR)-coated laser front facet with a reflectivity of R_f. An external reflector returns a fraction, R_{ext}, of the BAL output power back towards the laser.

The back-coupling to the semiconductor waveguide can be quantified by the factor $\Gamma_c = \eta_{\text{vert}} \eta_{\text{lat}}$, where η_{vert} and η_{lat} are the vertical (y) and lateral (x) coupling efficiencies, respectively. Due to strong vertical index guiding, η_{vert} can be expressed by the vertical overlap integral

$$\eta_{\text{vert}} = \frac{\left| \int E_i^+ E_e^- \, \mathrm{d}y \right|^2}{\int \left| E_i^+ \right|^2 \mathrm{d}y \int \left| E_e^- \right|^2 \mathrm{d}y}. \tag{2.2}$$

Because of highly dynamic filamentation present in the lateral direction of BALs, it is more meaningful to estimate η_{lat} by the overlap of the time-averaged near and far-fields of emitted and re-entering intensity, assuming suitable (filamentation-free) analytical functions for the corresponding distributions, as done e.g. in [34]. For matched feedback, Γ_c is unity.

The fraction of back-coupled light reentering the laser can be derived by considering the external cavity as a Fabry-Pérot etalon with mirror reflectivities R_f and R_e. Based on this approximation, optical feedback from such an external cavity can be reduced to a single reflection of E_i^+ at the front facet with an effective reflectivity R_{eff} (cf. Fig. 2.1, bottom). For the case of incoherent feedback, R_{eff} can then be derived from the phase-averaged reflectivity of a Fabry-Pérot etalon to [68]

$$R_{\text{eff}} = \frac{R_{\text{f}} + R_{\text{e}} - 2R_{\text{f}}R_{\text{e}}}{1 - R_{\text{f}}R_{\text{e}}}. \tag{2.3}$$

Thereby the total fraction of the front-facet output power coupled back into the waveguide is given by $R_{\text{eff}}\Gamma_{\text{c}}$.

This compound cavity model automatically reduces the investigation of optical feedback to a steady-state analysis where all dynamics caused by the delayed back-coupling of feedback light, such as sub-microsecond spatio-temporal fluctuations, are neglected.

For spectral stabilization, R_{e} is chosen to have a narrow-band peak at the desired stabilization wavelength λ_{s} and the front facet is AR coated ($R_{\text{f}} \ll R_{\text{e}}(\lambda_{\text{s}})$) so that, according to equation (2.3), $R_{\text{eff}} \approx R_{\text{e}}$ and the laser characteristics are dominated by the external cavity if $\lambda_{\text{p}} = \lambda_{\text{s}}$, with λ_{p} being the gain peak wavelength. In general, the threshold condition for lasing is that the total cavity losses consisting of internal absorption loss α_{i} and out-coupling loss $\alpha_{\text{m}} = -1/(2L_{\text{i}}) \ln (R_{\text{b}}R_{\text{eff}})$ are compensated by the modal gain g_{mod} averaged over the longitudinal (z) cavity length L_{i} [69]

$$\frac{1}{L_{\text{i}}} \int_0^{L_{\text{i}}} g_{\text{mod}} \, \mathrm{d}z = \alpha_{\text{i}} + \alpha_{\text{m}}. \tag{2.4}$$

To understand the integration bounds, note that the right-handed coordinate system used throughout this thesis has its origin in the QW at the center of the contact stripe at the back facet. In case of a mismatch of the gain peak wavelength λ_{p} and λ_{s}, the resulting lasing wavelength λ_0 aligns to either of them according to which of the corresponding operation states exhibits the lowest injection current that fulfills equation (2.4). This current minimum is the lasing threshold current I_{th}.

The presented analysis in this chapter targets investigating the feedback impact on BALs without dispersion of R_{e}. Therefore, the peak reflectivity of the external reflector $R_{\text{e}}(\lambda_{\text{s}})$ is virtually extended to the full spectral range, effectively treating it as a constant.

An important consequence of the concept of matched feedback is, in particular, that free-running and external-cavity operation are equal for steady-state analyses. Within this framework, investigations based on the variation of facet reflectivities are valid for both operation modes alike. Yet dynamical phenomena such as feedback-induced lateral intensity modulation have to be analyzed by using a true coupled cavity.

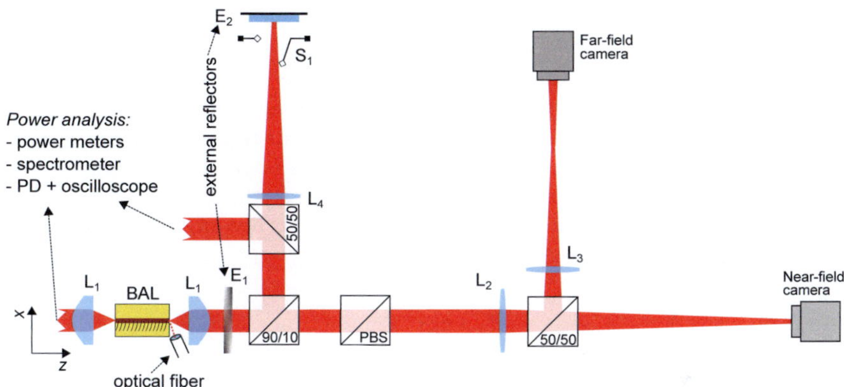

Fig. 2.2.: Experimental setup used for BAL characterization regarding beam quality and e-o performance and junction temperature for both output facets.

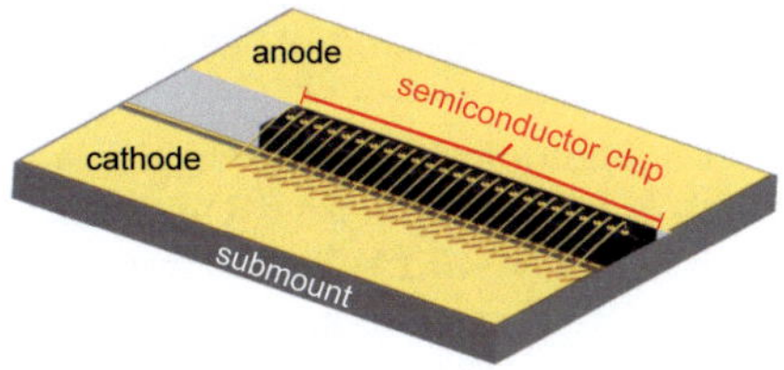

Fig. 2.3.: Exemplary assembly of a BAL on submount with electrical contact pads.

2.2. Experimental setup

The experimental setup used for the investigation of BAL lateral beam quality is sketched in Fig. 2.2. The devices under test (DUTs) are driven by a current source that allows CW and quasi continuous-wave (QCW) operation with various duty cycles as well as short-pulse mode. The assembly of a semiconductor chip on submount with contact pads is shown in Fig. 2.3. Via these contact pads, both the diode voltage is measured and the laser is supplied with injection current by two separate sets of electrical connections. For operation, the devices are clamped down to a water-cooled copper heat sink and are electrically probed by means of a mechanical fixture.

The intensity distribution on the facets, i.e. the near fields, are obtained by a telecentric imaging of the facets by lens L_1 and L_2 onto a CMOS camera. Here, the aspheric lens L_1 with a focal length of $f_{L_1} = 8\,\mathrm{mm}$ and numerical aperture of 0.5 is used for collimation. Afterwards, the transverse-electric (TE) polarized light is separated by a polarizing beam splitter (PBS) and imaged by lens L_2 (focal length $f_{L_2} = 300\,\text{mm}$) on the near-field camera chip with magnification of f_{L_2}/f_{L_1}.

For simultaneous evaluation of the far field, the light behind L_2 is split by a 50/50 beam splitter cube and is directed through lens L_3 (focal length $f_{L_3} = 200\,\text{mm}$) towards the far-field camera. The distances between L_2 and the two cameras amount to f_{L_2} and the far-field camera is positioned in the focal plane of L_3. By this arrangement, the lateral coordinate x at the position of the far-field camera is related to the lateral divergence angle θ of the laser, given in paraxial approximation by [23, 55]

$$\theta = x\frac{f_{L_2}}{f_{L_1}f_{L_3}}. \tag{2.5}$$

The exposure times of the near- and far-field cameras are set to $100\,\text{ms}$ and $15\,\text{ms}$, respectively. Moreover, the accuracies for the near- and far-field measurement amount to $\pm1\,\mu\text{m}$ and $\pm0.2°$, respectively.

The spatially and angle-resolved analysis of the emission is done for both facets consecutively by rotating the laser 180° around the vertical axis. In contrast, a spectral and time-domain analysis of the BAL power can be done simultaneously for both facets by means of optical spectrometers and an oscilloscope connected to a photodiode (PD). For single-side emitting DUTs, that by definition exhibit a high-reflection (HR)-coated back facet, the small fraction of light power (typically $<0.1\,\text{W}$) that transmits the HR coating is sufficient to measure the beam parameters.

The change of the junction temperature T_{jun} as a function of injection current I is derived by means of the thermal resistance R_{th} and the heat power P_{heat} which is the difference between electrical input and optical output power:

$$T_{\text{jun}}(I) - T_{\text{hs}} = \underbrace{\frac{\partial\lambda_0}{\partial P_{\text{heat}}}\left(\frac{\partial\lambda_0}{\partial T_{\text{hs}}}\right)^{-1}}_{=:R_{\text{th}}} P_{\text{heat}}(I). \tag{2.6}$$

Here, $\partial\lambda_0/\partial P_{\text{heat}}$ and $\partial\lambda_0/\partial T_{\text{hs}}$ are obtained from linear fits to the experimentally obtained shift of the centroid lasing wavelength λ_0. Whereas the slopes $\partial\lambda_0/\partial P_{\text{heat}}$ are derived for each device in regular CW operation, the value $\partial\lambda_0/\partial T_{\text{hs}} = (0.30 \pm 0.01)\,\text{nm}\,\text{K}^{-1}$ is obtained in short-pulse mode for a set of reference devices as a function of the heat sink temperature T_{hs}.

For external-cavity operation, two different types of cavity architectures are investigated by insertion of the external reflectors E_1 and E_2 referred to as 2f and 4f cavity, respectively, in the following. The 2f cavity is realized by inserting E_1 at a distance of $L_e = 2f_{L_1}$ behind the DUT. It is characterized by the re-entering light intensity being a delayed inverse self-image of the near field in both axis [70]. For investigation of the 2f cavity, the shutter S_1 in front of E_2 needs to be closed to avoid double-feedback. Instead, for the 4f cavity E_1 is not inserted and S_1 is opened. In this case the re-entering light is a self-image of the emitted field in both axis. This is achieved by the light being imaged onto E_2 by a telecentric lens arrangement of L_1 and L_4 (focal length $f_{L_4} = 250\,\text{mm}$). These two configurations are investigated since they provide the highest back-coupling efficiencies [70]. Since the 2f arrangements are potentially smaller and need less optical elements as compared to 4f configurations, they are preferred for small-footprint systems.

Apart from the analysis of the stimulated part of the BAL emission, the spontaneous emission at the front facet is also investigated to gain information about the carrier density distribution at the front facet. As illustrated in Fig. 2.2, light emitted under a lateral angle much larger than θ_{95}, i.e. $\theta \approx 60°$, is collected in the far field by an optical fiber and processed by an optical spectrometer. The result is the power spectral density $\tilde{P}_{\text{sp}}$ which is related to the net carrier density distribution $N(x)$ in the QW by the recombination rate of spontaneous emission R_{sp}, given by the relations [71]

$$R_{\text{sp}} = BN^2, \tag{2.7}$$

and

$$R_{\text{sp}} = a_\text{c} \int \frac{\tilde{P}_{\text{sp}}(\lambda, N)}{\lambda}\, \mathrm{d}\lambda. \tag{2.8}$$

Here, B is the spontaneous emission coefficient and the coupling factor a_c accounts for the fraction of the total spontaneous power that enters the optical fiber. Since an exact determination of a_c is difficult, experimental values for the carrier density derived from equations (2.7) and (2.8) should be given as multiples of a reference value such as at the threshold [72].

For the employed experimental arrangement in Fig. 2.2 only the laterally integrated recombination rate of spontaneous emission $S_{\text{sp}} = \int R_{\text{sp}}\, \mathrm{d}x$ can be evaluated because $\tilde{P}_{\text{sp}}$ is measured in the far field. Since S_{sp} is a strictly monotonic function of $N(x)$, it can be used to compare relative changes of the total carriers at the front facet to estimate, for example, the amount of lateral carrier accumulation [73]. Similar to the techniques presented in [73, 74], for the evaluation of S_{sp}, here the spectral range of stimulated emission is left out from the integral over wavelength λ in equation (2.8) by $\pm 10\,\text{nm}$ around the centroid λ_0.

2.3. Laser structure

For the evaluation of the lateral-longitudinal heat sources and successive temperature distribution, the three main loss mechanisms are Joule heating, free-

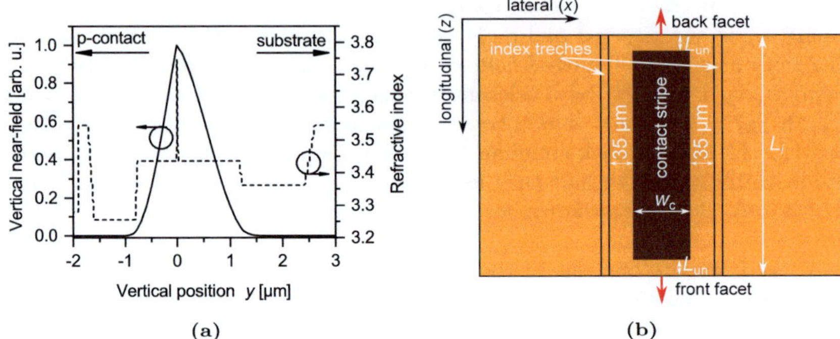

Fig. 2.4.: Refractive index and vertical near-field profile [75] (a) and lateral-longitudinal layout of the quasi gain-guided semiconductor chips under test (b). Here, laser light extraction is indicated by red arrows.

carrier absorption and non-radiative recombination. The spatial distribution of the corresponding heat source densities is primarily determined by the respective current-, photon- and carrier-density distributions. However, the contribution of each loss mechanism to the total waste heat is basically determined by the epitaxial layer structure.

Vertical structure

The epitaxial design of the BALs under investigation is grown by means of metal-organic chemical vapor deposition (MOCVD) on an n-doped GaAs substrate which is thinned to a thickness of 120 µm at the end of processing. The vertical structure is based on a compressively strained single InGaAs QW layer embedded in a slightly asymmetric 2-µm thick $Al_{0.19}Ga_{0.81}As$ waveguide. The thickness ratio of the n-doped to the p-doped waveguide part amounts to 60 : 40. This asymmetry ratio is kept for the thickness of the n- and p-doped cladding layers which are 1.2 µm and 0.8 µm, respectively. The cladding layers on each side of the QW are surrounded by a highly doped grading and cap layer which in total amount to 300 to 400 nm. The Ohmic contact at the p-side is achieved with a Ti-Pt-Au trilayer with a total thickness of 250 nm. The chip is soldered p-side down with AuSn on an AlN submount which is about 350 µm thick. The degree of TE polarization is >90 % for all investigated devices.

The vertical refractive index along with the corresponding near-field profile of the investigated vertical structure are displayed in Fig. 2.4a. The corresponding equivalent spot size d_{eq}, given by the ratio of the QW thickness d_{qw} over QW confinement factor Γ_{qw} [18], amounts to $d_{eq} = \frac{d_{qw}}{\Gamma_{qw}} = 1\,\mu m$. The corresponding far-field profile is of Gaussian shape and has a width of 44° (95 % p.c.).

The vertical layer structure, in particular the thickness, composition and doping of the individual layers, determines the fundamental electro-optical characteristics of the DUTs. Regarding optical losses, intra-band photon absorption by free carriers is the dominant mechanism in InGaAs-AlGaAs-based structures. Therefore it is assumed that the total internal loss α_i is due to free-carrier absorption. Hence, the internal absorption can be approximated by

$$\alpha_i = \sum_j \Gamma_j (\sigma_n N_{n,j} + \sigma_p N_{p,j}). \tag{2.9}$$

where Γ_j, $N_{n,j}$ and $N_{p,j}$ are the confinement factors, electron and hole densities, respectively, of the j epitaxial layers. For the quantitative evaluation of equation (2.9) in the presented analysis, $N_{n,j}$ and $N_{p,j}$ are chosen according to the respective nominal doping concentrations. σ_n and σ_p are the cross-sections for free-carrier absorption of electrons and holes, respectively. For the studied epitaxial structure, α_i is computed to $0.6\,\mathrm{cm}^{-1}$ assuming a threshold carrier density of $N = 2 \times 10^{24}\,\mathrm{m}^{-3}$ in the QW and $\sigma_{n,p} = 4, 12 \times 10^{-18}\mathrm{cm}^2$ [76]. Here, about one third of α_i is attributed to the QW, while the remaining two thirds are shared almost equally by the p- and n-doped sections.

Regarding Ohmic losses induced by a current I injected into the device, the series resistance R_s of the DUTs is derived from a linear fit to the device voltage U_0, which is given by

$$U_0 = U_{\mathrm{ph}} + \Delta U(N) + R_s I. \tag{2.10}$$

The photon voltage U_{ph} is defined by $hc_0/(\lambda_0 e)$ with h, c_0 and e being Planck's constant, the vacuum speed of light and the elementary charge, respectively. The parameter ΔU is referred to as quantum defect voltage.

Lateral-longitudinal structure

The lateral-longitudinal layout of the DUTs is schematically depicted in Fig. 2.4b. The length of the rectangular contact stripe is by L_{un} shorter than the resonator on either side of the emitter, to prevent direct electrical pumping of the sections adjacent to the facets and thus carrier recombination at the surface. These non-pumped sections have a length of $L_{\mathrm{un}} = 30\,\mu\mathrm{m}$. The investigated lateral-longitudinal geometries exhibits contact widths w_c of 100, 120, 130 and 140 μm and internal cavity lengths of 4 mm and 5 mm. The variety of resonator geometries is of minor importance for the physical effects studied but rather attributed to their logistical availability.

In terms of lateral chip designs, two types of lateral structuring are examined: The first type is a quasi gain-guided approach where in the adjacent

$35\,\mu$m-wide regions to the contact area there are no built-in index steps so that the electric field is essentially gain-guided. Adjacent deep index trenches are employed only to avoid lateral light leakage in case of laterally misaligned optical feedback. The corresponding effective refractive index change [77] Δn_0 ranges from 10^{-3} to 10^{-2}.

The second type is an ion-implanted design. The main idea behind ion-implantation is that the bombarded areas become insulating. Therefore, it is frequently used as a means of suppressing lateral current spreading. It has been reported that implantation through the QW leads to a reduced lateral carrier accumulation and a lower θ_{95}. Such a reduction of lateral carrier accumulation is attributed, besides a reduced current spreading, to a higher rate of non-radiative recombination by the defects created in the QW [73]. However, this beneficial impact of deep ion-implantation can be accompanied by a reduced power slope efficiency.

The implantation design studied in this thesis consists of a three-step proton bombardment at a distance of $5\,\mu$m beside the contact stripe, which leads to a proton density of about $1 \times 10^{24}\,\mathrm{m}^{-3}$ at the position of the QW. Concerning device simulation, a reduced power slope for the ion-implanted design can be accounted for by an increased absorption coefficient compared to the gain-guided design.

2.4. Simulation of lateral emission characteristics of single-side emitting BALs

In this section, the simulated lateral-longitudinal temperature profile $T(x, z)$ of BALs is examined with respect to its impact on beam quality and lifetime. The analysis is conducted for BALs emitting in the wavelength range from 960 to 980 nm but is generally valid for the InGaAs/AlGasAs material system.

In the course of this thesis, the electro-thermo-optical properties of BALs are simulated numerically with two different software packages, namely *BALaser* and *SEMSIS*. The reason to use two simulation codes can be attributed to their temporal availability. At first, the BALaser tool developed externally by the Weierstrass Institute for Applied Analysis and Stochastics in Berlin [78] was used until the more flexible code of SEMSIS software became available within the TRUMPF Group. Whereas both programs have their individual advantages, they perform similarly concerning the lateral beam parameters of BALs, as will be shown in the following. The physical background of these two programs is briefly introduced in the next subsection with the focus on their respective thermal solvers.

2.4.1. Simulation tools

BALaser

The BALaser program is based on a time-dependent traveling wave model that describes the slowly varying amplitudes of the electric field $E^{\pm}(x, z, t)$ for TE polarization by vertically projected, paraxial parabolic equations [79–81]

$$\frac{n_{\mathrm{g}}}{c_0}\partial_t E^{\pm} \pm \partial_z E^{\pm} + \frac{i}{2k_0\bar{n}}\partial_{xx}E^{\pm} = -i\beta E^{\pm} + \hat{D}(t)E^{\pm} + F_{\mathrm{sp}}^{\pm}, \qquad (2.11)$$

which are coupled to a diffusion equation for the net carrier density in the QW $N(x, z, t)$ of the form [51]

$$\partial_t N = D_N \partial_{xx} N + \frac{J}{ed_{\mathrm{qw}}} - R_{\mathrm{nr,sp}}(N) - R_{\mathrm{stim}}(N, E^{\pm}). \qquad (2.12)$$

The parameters n_{g}, $\bar{n}$, D_N and J are the group refractive index, a reference refractive index, diffusivity, and injection current density, respectively. Further, $R_{\mathrm{nr,sp}}$, R_{stim} and $F_{\mathrm{sp}}^{\pm}$ denote the non-radiative and spontaneous radiative recombination rate, the stimulated emission rate, and the contribution of spontaneous emission to the optical field, respectively. The equations (2.11) and (2.12) are coupled to ordinary differential equations for the material polarization via the term $\hat{D}(t)E^{\pm}$ to account for gain dispersion [81,82]. The boundary conditions for equation (2.11) at the facets read

$$E^{+}(z = 0) = \sqrt{R_{\mathrm{b}}}E^{-}(z = 0) \quad \text{and}$$
$$E^{-}(z = L_{\mathrm{i}}) = \sqrt{R_{\mathrm{f}}}E^{+}(z = L_{\mathrm{i}}). \qquad (2.13)$$

Finally, the propagation factor β consists of the sum of the effective refractive index changes as well as gain and absorption loss [79]:

$$\beta = k_0(\Delta n_0 + \Delta n_N + \Delta n_T) + \frac{i}{2}(g_{\mathrm{mod}} - \alpha_{\mathrm{i}}), \qquad (2.14)$$

with $k_0 = 2\pi/\lambda_0$. Regarding the modal gain, a logarithmic model of the form $g_{\mathrm{mod}} = g_0 \ln(N/N_{\mathrm{tr}})$ is assumed with N_{tr} and g_0 being the transparency carrier density and modal gain coefficient, respectively. Changes of the effective refractive index [51] due to temperature Δn_T and carriers Δn_N are included via

$$\Delta n_T = \frac{\partial n}{\partial T}(T(x, z) - T_{\mathrm{hs}}) \qquad (2.15)$$

and

$$\Delta n_N = -\sqrt{n' N(x, z, t)}, \qquad (2.16)$$

where $T(x, z) := T(x, y = 0, z)$ and n' are the temperature profile at the QW position and the differential index, respectively.

At the time of use (2016 – 2017), BALaser was a sole electro-optical solver for $E^{\pm}(x, z, t)$ and $N(x, z, t)$ using a splitting scheme of finite differences and Fourier beam propagation [80]. It was part of the present thesis to couple the results of BALaser iteratively to a thermal solver based on the Finite-Element Method (FEM) to obtain Δn_T. With this solver, $T(x, y, z)$ is approximated by solving the steady-state heat conduction equation [52]

$$\nabla_{(x,y)} \left[\kappa(x, y) \nabla_{(x,y)} T(x, y, z_0) \right] = -H(x, y, z_0) \tag{2.17}$$

in the (x, y)-plane at several sampling points z_0 along the longitudinal axis. The total heat source density $H(x, y, z_0)$ is considered a sum of the three main heat source densities attributed to optical absorption (H_{abs}), non-radiative recombination (H_{nr}) as well as to carrier injection which causes Joule and quantum defect heating ($H_{\mathrm{j+d}}$):

$$H(x, y, z) = H_{\mathrm{abs}}(x, y, z) + H_{\mathrm{nr}}(x, y, z) + H_{\mathrm{j+d}}(x, y, z) \tag{2.18}$$

To solve equation (2.17) along with (2.18), a simplified 6-layer vertical layer system is considered consisting of n-contact, substrate, one n- and p-doped layer each, solder and submount. The corresponding thermal conductivities are chosen similar to those reported in [52,83]. Within this framework, all heat source densities are distributed vertically homogeneously over the p-layer with a thickness of $d_{\mathrm{p}} = 2\,\mu\mathrm{m}$, which is set equal to the thickness of the vertical waveguide. This approximation is valid since changes due to an epitaxial layer-resolved heat source density are small due to the smoothing properties of the heat equation. Also in the lateral direction a simplified geometry is used with the heat power densities being piecewise constant. Thus H_{abs}, H_{nr} and $H_{\mathrm{j+d}}$ are approximated by

$$H_{\mathrm{abs}}(x, y, z) = \begin{cases} \alpha_i S(z) \frac{d_{\mathrm{eq}}}{d_{\mathrm{p}}} & \text{for } -\frac{w_{95}(z)}{2} \leq x \leq \frac{w_{95}(z)}{2} \wedge y \in \text{p-layer} \\ 0 & \text{else,} \end{cases} \tag{2.19}$$

$$H_{\mathrm{nr}}(x, y, z) = \begin{cases} eU_{\mathrm{ph}} R_{\mathrm{nr,sp}}(N(z)) \frac{d_{\mathrm{qw}}}{d_{\mathrm{p}}} & \text{for } -\frac{w_c}{2} \leq x \leq \frac{w_c}{2} \wedge y \in \text{p-layer} \\ 0 & \text{else} \end{cases} \tag{2.20}$$

and

$$H_{\mathrm{j+d}}(x, y, z) = \begin{cases} (U_0 - U_{\mathrm{ph}}) J(z)/d_{\mathrm{p}} & \text{for } -\frac{w_c}{2} \leq x \leq \frac{w_c}{2} \wedge y \in \text{p-layer} \\ 0 & \text{else ,} \end{cases} \tag{2.21}$$

respectively.

Here, the quantities $S(x, z) = eU_{\mathrm{ph}} \frac{c_0}{n_{\mathrm{g}}} \frac{d_{\mathrm{qw}}}{d_{\mathrm{eq}}} \left\langle \left| E^+(x, z, t) \right|^2 + \left| E^-(x, z, t) \right|^2 \right\rangle$ and $N(x, z) = \langle N(x, z, t) \rangle$ represent the time-averaged internal intensity and net carrier density, respectively. Hereby, a period of 1 ns turned out to be sufficient for averaging to obtain stationary distributions. Note that $S(x, z)$ is referenced to d_{eq} in vertical direction, meaning that the corresponding power value can be obtained via $d_{\mathrm{eq}} \int S(x, z)\,\mathrm{d}x$. Furthermore, in the equations (2.19)-(2.21), $S(z)$ and $N(z)$ are calculated by laterally averaging $S(x, z)$ and $N(x, z)$ over $w_{95}(z)$ and w_c, respectively, with $w_{95}(z)$ being the lateral width of $S(x, z)$ along the resonator axis including 95 % p.c.

The rate of non-radiative and spontaneous radiative recombination reads $R_{\mathrm{nr,sp}}(N) = AN + BN^2 + CN^3$ with A and C being the coefficient for Shockley-Read-Hall and Auger recombination, respectively [79]. For the spontaneous and Auger processes, the recombination rates are proportional to the carrier densities of the involved electron and hole states in the conduction and valence bands, respectively [53]: one electron and hole each for spontaneous emission and two electrons + one hole or one electron + two holes for Auger recombination. For the Shockley-Read-Hall process, however, recombination via a midgap defect state takes place so that the corresponding rate is proportional to N [84].

Moreover, the total injection current is distributed along the contact stripe according to the Fermi voltage U_{F} via the current density [51]

$$J(z) = \frac{U_0 - U_{\mathrm{F}}(N(z))}{R_s w_c L_i}. \tag{2.22}$$

The Fermi voltage is given by the energy difference of the quasi-Fermi levels in the QW divided by the elementary charge. U_{F} is fit to microscopic band calculations [85] via the function $U_{\mathrm{F}}(N) = a_{\mathrm{F}} + b_{\mathrm{F}}(N\mathrm{m}^3)^{c_{\mathrm{F}}}$, where a_{F} is afterwards adjusted in order to assure $w_c \int J(z)\,\mathrm{d}z = I$ holds. Outside the contact stripe, it is assumed that $I = 0$, meaning that lateral current spreading is neglected.

The solution $T(x, y, z_0)$ of equation (2.17) is calculated at 80 equally spaced locations z_0 along the cavity. For the boundary condition to the heat sink at the bottom of the submount, a Robin boundary condition of the form $\partial_y T = h_{\mathrm{hs}}(T - T_{\mathrm{hs}})$ is applied [86]. Here, the conduction coefficient h_{hs} is used as fitting parameter. It is chosen such that the mean QW temperature $\overline{T}_{\mathrm{QW}} = 1/(w_c L_i) \iint T(x, z)\,\mathrm{d}x\,\mathrm{d}z$, averaged over the contact area, matches the experimental junction temperature T_{jun} obtained from equation (2.6). All other boundaries are considered adiabatic.

The thermal solver is iteratively coupled to BALaser via equation (2.15) and parameters which are considered dependent on temperature. The corresponding numerical values used for the extended BALaser package are listed in the appendix A (cf. Tab. A.1).

In conclusion, an electro-thermo-optical analysis based on the presented extended BALaser package for a given DUT requires, beside these material parameters, experimental input of $U_0(I)$ and h_{hs}.

SEMSIS

The software package called SEMSIS was originally developed at the Fraunhofer Institute for Laser Technology ILT in Aachen, Germany. SEMSIS provides a frequency-based electro-optical solver for the parabolic carrier diffusion (stationary form of equation (2.12)) and extended Helmholtz equation in the lateral-longitudinal plane that are solved with a wide-angle finite difference approach [52, 87]. The electro-optical solver is coupled to a three-dimensional thermal solver which, in contrast to the BALaser extension, includes longitudinal heat flow and resolves the actual vertical layer structure down to the QW dimension. Therefore, SEMSIS is advantageous for analyses of close-to two-dimensional heat distributions located in either the lateral-longitudinal or the lateral-vertical plane. The configuration of SEMSIS used for this thesis is summarized below.

The two-dimensional electro-optical solver employs a monochromatic travelling-wave model in connection with the vertically projected effective index model of equation (2.14), but with a linear expression concerning carrier-induced index changes, given by $\Delta n_N^{\mathrm{sem}} = -n'N(x,z)$. The current injection is set stepwise constant with $I = 0$ beside the contact area. In particular, lateral current spreading effects are disregarded by this choice. The output of the electro-optical solver is the internal intensity and excess carrier density which are denoted, just as in the case of the BALaser package, by $S(x,z)$ and $N(x,z)$, respectively. These quantities are obtained by averaging over a variable number of round-trips and are used as input for the thermal solver.

To model the vertical distribution of the heat sources, SEMSIS additionally includes a one-dimensional electrical solver [87]. This solver is based on a drift-diffusion model for electron and hole transport and is able to compute a steady-state solution of the vertical electron and hole densities $N_{\mathrm{n}}(y)$ and $N_{\mathrm{p}}(y)$, respectively, for an applied U_0.

The absorption heat power density for the SEMSIS package $H_{\mathrm{abs}}^{\mathrm{sem}}$ is:

$$H_{\mathrm{abs}}^{\mathrm{sem}}(x,y,z) = \begin{cases} \alpha_0 S_{\mathrm{vert}}(y) d_{\mathrm{eq}} S(x,z) & \text{for } y \neq 0 \\ \Gamma_{\mathrm{qw}}(\sigma_{\mathrm{n}} + \sigma_{\mathrm{p}}) N(x,z) d_{\mathrm{eq}} S(x,z) & \text{for } y = 0. \end{cases} \tag{2.23}$$

Here, the absorption in the waveguide, represented by the coefficient α_0, is considered constant whereas in the QW it depends on $N(x,z)$. Moreover, $S_{\mathrm{vert}}(y)$ denotes the vertical near-field distribution shown in Fig. 2.4a, normalized so that $\int S_{\mathrm{vert}}(y)\,\mathrm{d}y = 1$. Thereby absorption according to the vertical mode profile is considered whereas the vertical doping profile is neglected in equation (2.23).

Furthermore, the heat power density due to non-radiative recombination is used like in the BALaser package (equation (2.18)), but vertically confined to the QW region only and is spatially resolved both in lateral and longitudinal direction. The remaining heat power which adds up to P_{heat} is assumed to be due to Joule heating which is distributed homogeneously in the (x, z)-plane and in the vertical direction according to the contribution of the resistance R_j of each layer j to the total series resistance $\sum_j R_j$, which is evaluated based on the results of the electrical solver. With these heat source distributions, $T(x, y, z)$ is calculated by a three-dimensional form of equation (2.17) with the same boundary conditions for the thermal solver used in combination with BALaser. The thermal conductivities used for the individual epitaxial layers are listed in [87]. By the solution of $T(x, y, z)$, the thermal refractive index (2.15) and all temperature-dependent parameters are iteratively updated for the next iteration of the electro-optical solver. This iteration process is repeated at each 20th round-trip for 200 round-trips in total. The parameters used for the electro-optical simulations are listed in the appendix A, Tab. A.1. The parameter database behind SEMSIS (cf. [87]) is in part based on slightly different microscopic models as used for BALaser (cf. [85]), which explains differences in the simulations parameters employed for the two tools.

For both simulation packages the back and front-side near-field distribution are given by $S(x, z = 0)\frac{1-R_b}{1+R_b}$ and $S(x, z = L_i)\frac{1-R_f}{1+R_f}$, respectively. The far field profiles at the back and front facet are $\propto \left\langle \left| \mathscr{F} \left\{ E^{\pm}(x, z = 0, L_i) \right\} \right|^2 \right\rangle$, where $\mathscr{F}$ denotes the Fourier transform. Moreover, for both tools intrinsic surface heat sources such as enhanced recombination or absorption at the facets are neglected since the employed laser structure employs non-pumped mirrors and facet passivation with a high-bandgap material.

Due to the nonlinear coupling of the electric field and carrier density (cf. equations (2.11) and (2.12)), there is no analytical solution for the field propagation in BALs which complicates the validation of numerical tools. Therefore, in the interim phase when both tools were available, a benchmark between them was conducted to ensure consistency between the two different numerical approaches with reference to the lateral beam parameters.

In Fig. 2.5 the near- and far-field distribution of a BAL with $w_c = 90\,\mu\text{m}$ is simulated for exactly the same parameter set and lateral-longitudinal temperature distribution with BALaser and SEMSIS. For this comparison, both programs use equation (2.16) to account for carrier-induced changes of the effective refractive index. It can be seen in Fig. 2.5 that the edges of near- and far-field distributions from both tools overlap. Thus the beam parameters w_{95} and θ_{95} are almost equal. This observation means that the employed simulation codes produce consistent results concerning the lateral beam parameters. Moreover both tools produce a similar number of filaments, but the corre-

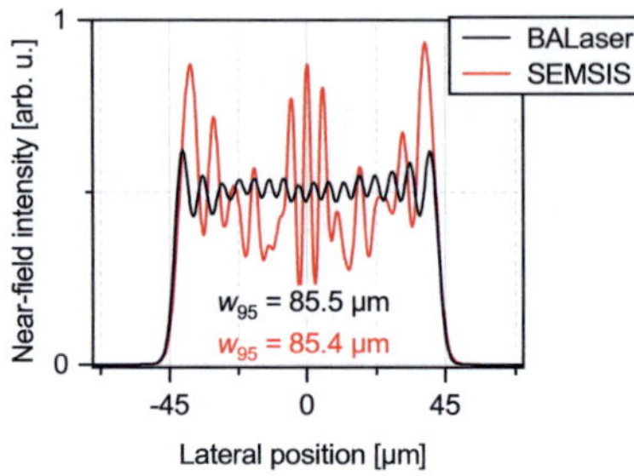 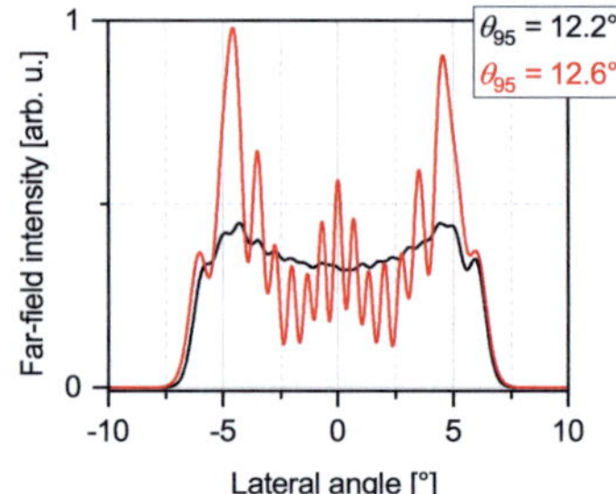

Fig. 2.5.: Comparison of simulations codes for the same parameter set: While the widths of the near- and far field are almost the same, their filmentation behavior differs strongly, which is mainly due to a different way of averaging.

sponding filamentation amplitude is much more pronounced for the SEMSIS result. This deviation is due to the different way of averaging to calculate $S(x, z)$: While SEMSIS averages the intensity distributions of 50 consecutive round trips, for BALaser an averaging of 1920 equidistant time steps within 1 ns is chosen. As a consequence, the BALaser output is more smoothed out than SEMSIS results.

2.4.2. Lateral-longitudinal temperature profile

This subsection introduces the global trends of $T(x, z)$ as a function of injection current, simulated by the extended BALaser package for an ion-implanted BAL with $w_c = 120\,\mu m$ and $L_i = 4\,mm$ at a heat sink temperature of $29\,°C$. The facet reflectivities are $R_f = 2\,\%$ and $R_b = 96\,\%$. This DUT represents a frequently used coating combination for single-side emitting BALs.

To avoid thermal lensing in the experimental setup, the device is operated in QCW mode with pulses of $1\,ms$ and $37\,\%$ duty cycle (d.c.) which results in a thermal resistance of $R_{th} = 3.1\,K\,W^{-1}$. Since longitudinal heat flow is disregarded in these calculations and the discretization step size in the z-direction is $50\,\mu m$, the thermal effect of a non-pumped section in the same order of magnitude cannot be resolved properly and hence is neglected in these calculations $(L_{un} = 0)$.

In Fig. 2.6a the corresponding light-current (L-I) curve is depicted. The simulation is able to reproduce the experimental values with a maximum relative deviation of less than $7\,\%$. Moreover, Fig. 2.6a illustrates the fraction of the optical power P_{opt} and all dissipated power portions P_{abs}, P_{nr}, P_{j+d} relative to the total injected power $U_0 I$ according to conservation of energy

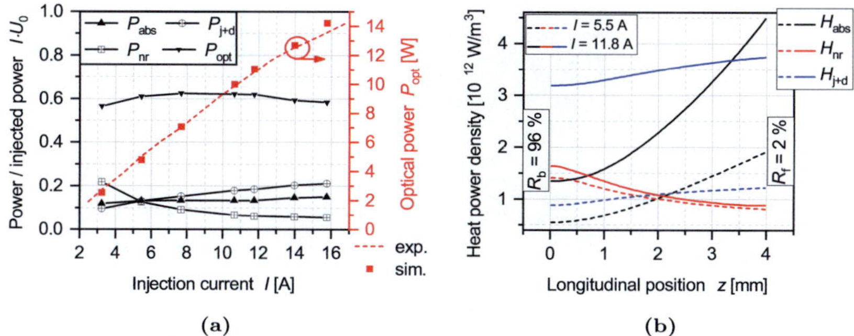

Fig. 2.6.: Power relations for a DUT with $w_c = 120\,\mu\text{m}$ and $L_i = 4\,\text{mm}$ simulated with the extended BALaser software: Optical and dissipated power fractions as well as absolute optical power vs. current (a) and longitudinal distribution of the different heat sources for 5.5 A and 11.8 A (b).

$$U_0 I = P_{\text{opt}} + P_{\text{abs}} + P_{\text{nr}} + P_{\text{j+d}}. \tag{2.24}$$

Each term representing a dissipated power in equation (2.24) is given by the volume integral of the respective heat power density defined in the equations (2.19)–(2.21). The term $\eta_{\text{eo}} = P_{\text{opt}}/(U_0 I)$ denotes the e-o efficiency.

The Fig. 2.6a shows that above 6 A the relative power contribution of $P_{\text{j+d}}$ is strongest and increases nearly linearly, which is because Joule heating scales with I^2. In contrast, the contribution of P_{abs} to $U_0 I$ remains nearly constant and the fraction of P_{nr} decreases with increasing I in the investigated current interval. Consequently, $\overline{T}_{\text{QW}}$ is influenced strongest by $P_{\text{j+d}}$, the longitudinal homogeneity of $T(x,z)$, however, is not – as can be seen in Fig. 2.6b. There, the heat power densities are plotted over the resonator axis for currents of 5.5 A and 11.8 A.

According to the equations (2.20) and (2.21) H_{nr} and $H_{\text{j+d}}$ are monotonic functions of $N(z)$ and H_{abs} is proportional to the total internal intensity $S(z)$. By implication, the profiles of $S(z)$ and $N(z)$ determine the distributions of the corresponding heat power densities: As it is known from one-dimensional calculations neglecting lateral effects [69, 88, 89], the strong asymmetry of R_{f} and R_{b} induces $S(z)$ growing monotonically from back to front facet. Additionally, the growth rate $\partial\left[S(L_i) - S(0)\right]/\partial I$ can be approximated to $\sim$70 % of that of the optical power $\partial P_{\text{opt}}/\partial I$ (cf. Fig. 2.6a) by the formulas given in [66]. In contrast, as a consequence of gain saturation, $N(z)$ decreases monotonically from back to front facet which is referred to as (long-range) longitudinal spatial hole burning (SHB) [66]. And, in particular, $N(z)$ grows much weaker with increasing current due to gain clamping at threshold [66].

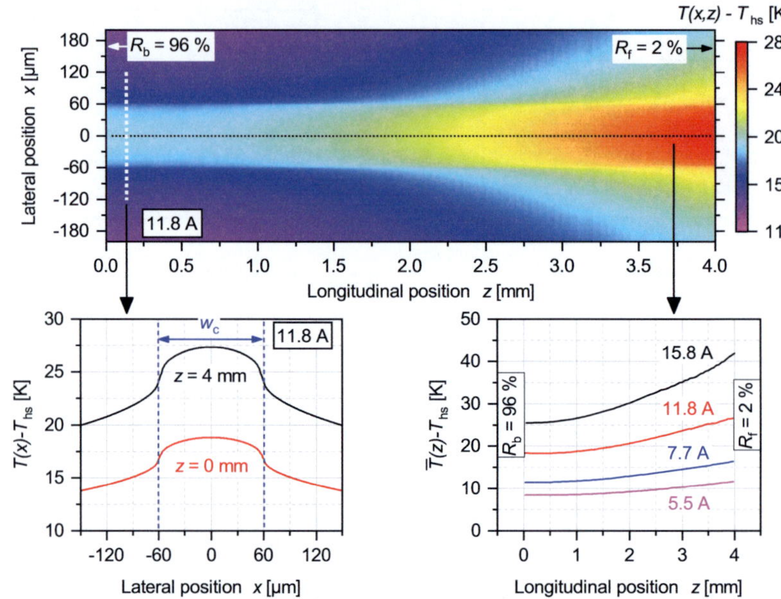

Fig. 2.7.: Lateral-longitudinal temperature distribution at the position of the QW for a DUT with $w_c = 120\,\mu\mathrm{m}$ operated QCW at current of 11.8 A. The data is simulated with the extended BALaser software. Lateral (bottom left-hand side) and longitudinal cross sections (bottom right-hand side) show that both the temperature differences from the middle of the contact stripe to the edges as well as from back to front facet rise with increasing current. Published in excerpts in [90].

These are the dominant effects that cause the growth rates $\partial\,|H_k(L_i) - H_k(0)|\,/\partial I$ to be larger for $k \in \{\mathrm{abs}\}$ than for $k \in \{\mathrm{nr, j+d}\}$, as exemplarily illustrated in Fig. 2.6b. In conclusion, for single-side emitting high-power BALs operating at high currents, Joule heating dominates the absolute device temperature, whereas optical absorption has dominant impact on the temperature difference between the front and back facet.

The full solution of $T(x, z) - T_{\mathrm{hs}}$ is shown in Fig. 2.7 for a current of 11.8 A along with lateral (bottom left) and longitudinal cross sections (bottom right). While the lateral cross sections show $T(x, z_0 = 0, 4\,\mathrm{mm})$, the longitudinal cross sections are laterally averaged over w_c, indicated by $\overline{T}(z)$. Since the lateral distributions of the heat source densities in equation (2.18) are chosen symmetrically stepwise constant, the lateral cross sections are axis-symmetric.

Similar to comparable simulations [23, 91], these lateral cross sections exhibit pronounced peaks at the contact stripe that show highest drop in the vicinity of the contact edges, thus indicating the borders of main optical confinement. The simulated longitudinal cross sections show an increasing temperature difference between front and back facet with increasing current due to optical absorption. Neglecting longitudinal heat flow manifests itself by the fact that $\overline{T}(z)$ maintains the profile of $H(z)$. Incorporating heat flow in the z-direction would smooth the profiles resulting in a reduced temperature difference between the facets, but the monotony of curves would be conserved [49] – which is most important for the following analysis of the lateral beam parameters of this DUT.

2.4.3. Impact of longitudinal temperature profile on lateral emission characteristics

In Fig. 2.8 the experimental lateral near- and far-field widths at the back- and front facets are compared to the corresponding simulation data with and without longitudinal temperature variation (LTV). For the longitudinally isothermal data, equation (2.17) is evaluated at a single sampling point in z-direction, thereby averaging $N(z)$ and $S(z)$ over the resonator length. Thus, the obtained distribution $T(x)$ is present isothermally along the z-axis. In contrast, for the case with LTV the same simulation conditions as in Fig. 2.7 are chosen.

At first the experimental data are compared with the simulation data with LTV. In Fig. 2.8, it can be seen that the experimental profiles for both near- and far-field widths can be reproduced by the simulation with reasonable qualitative and quantitative agreement. It is striking in Fig. 2.8a that the near fields at both facets are at nearly the same level up to a current of $\sim$12 A, where they begin to diverge, with the front-facet and the back-facet near-fields shrinking and blooming, respectively, by up to $\sim$20 % with increasing current. A similar behavior, but not as pronounced, can be observed for the far-field widths: Up to a current of $\sim$12 A, the values for θ_{95} at both facets rise with increasing current with the back-facet far field being slightly larger. From $\sim$12 A towards higher currents, the slope $\partial\theta_{95}/\partial I$ at the front facet increases, so that the curves intersect and diverge.

In sum, at high injection currents, a near-field narrowing is accompanied by an additional blooming of the far-field at the front-facet and vice versa at the laser back. This behavior, however, is not visible for the case without LTV. Instead, the widths of near- and far field basically overlap at both facets for the whole investigated current range: The widths of the near field stay nearly constant and the divergences of the far fields increase linearly with rising current. This fundamentally different behavior between simulation with and without LTV has essential consequences listed in the following:

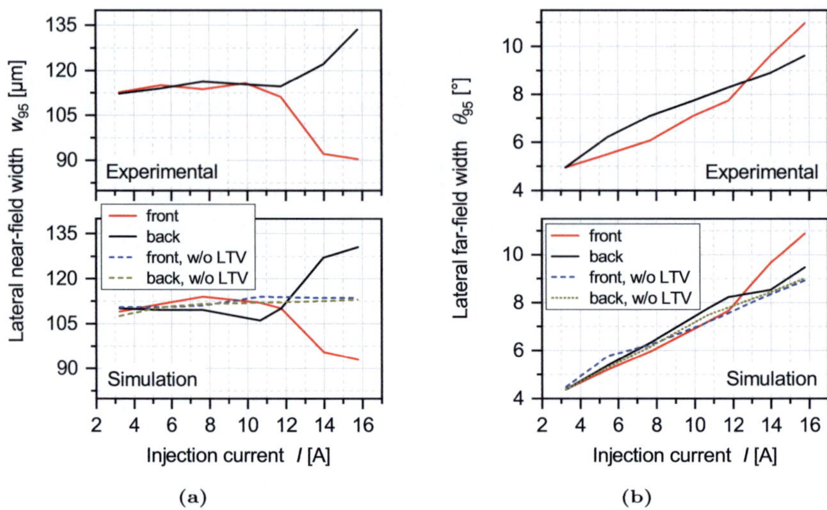

Fig. 2.8.: Comparison of experimental and simulated near- (a) and far-field widths (b) at back- and front facets for an ion-implanted DUT with $w_c = 120\,\mu m$ and $L_i = 4\,mm$ operated QCW with $37\,\%$ d.c. The simulated data is provided for operation with LTV, as illustrated in Fig. 2.7, and without LTV, i.e. isothermal operation in the longitudinal direction. Published in excerpts in [90].

- The experimentally observed differences between the front- and back-facet beam parameters of gain-guided single-side emitting BALs in the high current range can be explained by their longitudinal asymmetry of the internal temperature distribution: Lateral thermal lensing that increases from the back to the front of the laser causes the front facet near field to narrow and the corresponding far field to bloom additionally. Furthermore, LTV causes the carrier lensing to support the asymmetry between front- and back-side near field. For example, if carrier lensing is neglected (not shown in Fig. 2.8), i.e. $\Delta n_N = 0$, the difference of w_{95} between back- and front facet is reduced by about $15\,\%$ for the highest current value investigated.

- The reduction factor r_{nf} given by

$$r_{nf} = \frac{\min\left(w_{95}(I)\right)}{w_c} \tag{2.25}$$

is introduced in this thesis to quantify near-field narrowing with reference to the contact stripe width w_c. Here, r_{nf} denotes the intensity ratio of two idealized identical lasers emitting laterally rectangular near-field distributions, one with width $\min\left(w_{95}(I)\right)$ and the other with width w_c at the same output power.

With respect to laser reliability, r_nf can be used to estimate the lifetime change due to a reduction of the lateral near-field size: Assuming that laser lifetime can be described by an Arrhenius equation [63, 92, 93], a reduction of $t_\mathrm{CO(M)D}$ by sole geometrical change of the near-field scales exponentially with r_nf. Moreover, as can be derived from the comparison in Fig. 2.8a, a homogenization of the longitudinal temperature profile can be applied to prevent such a lifetime reduction.

- Thermally induced far-field blooming is known to be a significant contribution to an increasing $\mathrm{BPP}_\mathrm{lat}$ with increasing optical power [21, 65]. In Fig. 2.8b, the thermal and carrier-induced contribution to the raise of θ_{95} are superimposed in the simulated curves without LTV. The change of the blooming rate at $\sim 12\,\mathrm{A}$ due to LTV suggests that also thermal far-field blooming can be subdivided into two contributions – one due to longitudinally homogeneous thermal lensing and one due to LTV.

- For the front-facet data presented in Fig. 2.8, an evaluation of $\mathrm{BPP}_\mathrm{lat}$ is shown in Fig. 2.9. There, it can be seen that LTV leaves the front-facet $\mathrm{BPP}_\mathrm{lat}$, which contains the main optical power, unchanged, being a nearly linear function of I for the investigated current range. To check the general validity of this observation, $\mathrm{BPP}_\mathrm{lat}$ is investigated here for a set of (artificial) linear chirps g_c of $n_T(x, z)$ in the z-direction.
A lateral distribution $n_T^0(x)$ for $I = 14.5\,\mathrm{A}$ is chosen fixed at the center of the cavity. Then $n_T(x, z)$ is longitudinally chirped according to

$$n_T(x, z) = n_T^0(x) + g_\mathrm{c} \cdot n_T^0(x) \cdot (z - L_\mathrm{i}/2), \qquad (2.26)$$

with the slopes g_c varied between $[-0.9, 0.9]$. With this, it is assured that $n_T(x, z) > 0$ and $\iint n_T(x, z)\,\mathrm{d}x\,\mathrm{d}z = \text{constant}$. The latter condition is in good approximation equivalent to a constant average device temperature $\overline{T}_\mathrm{QW}$.

The normalized results of w_{95}, θ_{95} and $\mathrm{BPP}_\mathrm{lat}$ at the front facet are shown in Fig. 2.10 as a function of g_c. Here, the case without LTV ($g_\mathrm{c} = 0$) is chosen as a reference. For $|g_\mathrm{c}| < 0.7$, the trend that if one of the parameters w_{95} and θ_{95} increases, the other decreases and vice versa, is visible. In this parameter region, $\mathrm{BPP}_\mathrm{lat}$ remains within $<10\,\%$ change of the reference value. In contrast, for the outer positive value of g_c investigated, near-field narrowing saturates whereas far-field blooming continues. This behavior indicates a possible advantage of a longitudinally homogeneous temperature profile in terms of $\mathrm{BPP}_\mathrm{lat}$ for the highest drive currents.

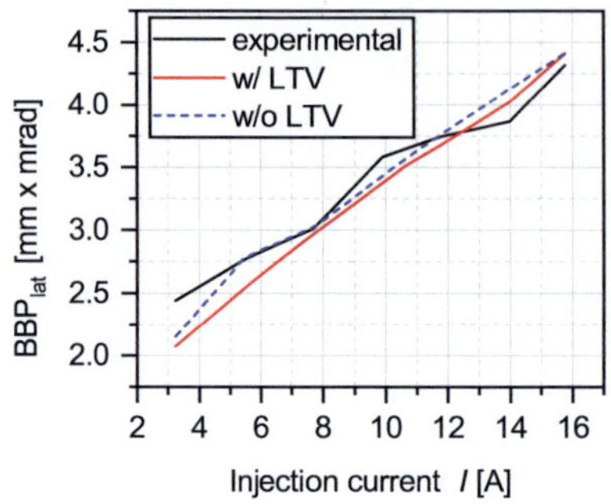

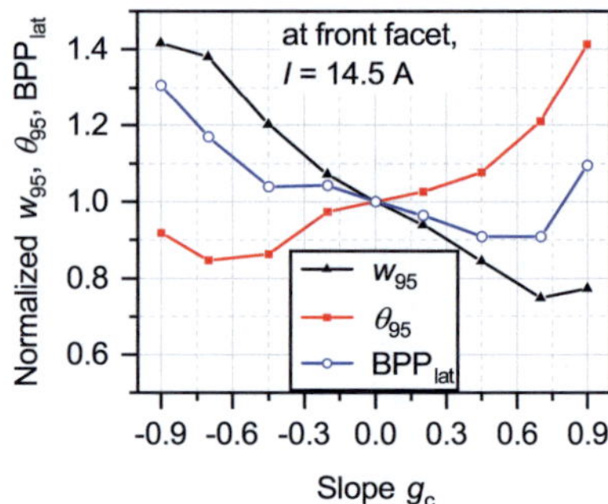

Fig. 2.9.: BPP_{lat} for the front-facet data of Fig. 2.8, [75].

Fig. 2.10.: Variation of longitudinal slope g_c (cf. equation (2.26)) of thermal effective refractive index change.

So far, the impact of LTV in BALs has been evaluated only for the near and far field, hence for the emission parameters that are accessible both experimentally and in simulation via the facets. The full two-dimensional information of $S(x, z)$ and $N(x, z)$, however, have the capability to illustrate the phenomena at the facets in more detail. Therefore, in Fig. 2.11, $n_T(x, z)$, $S(x, z)$ and $N(x, z)$ are shown for the highest current investigated in Fig. 2.8 where the impact of LTV is most evolved.

The comparison of $S(x, z)$ with and without LTV (middle row in Fig. 2.11) reveals that LTV causes the intensity to decrease successively towards the laser front, starting at about halfway along the cavity. As a result, LTV causes a down-tapering of the waveguide based on lateral thermal lensing. Once triggered, this process amplifies itself, since the absorption heat is distributed over a successively decreasing lateral stripe width. As a consequence of such a down-tapering, there are electrically pumped regions at the contact edges in the vicinity of the front facet where the intensity is so low that the carrier density exceeds the threshold values by several times (indicated red-colored area in Fig. 2.11, bottom left-hand side). This effect can be considered as increased lateral spatial hole-burning, which entails an increased non-radiative recombination lowering the internal differential efficiency

$$\eta_0 = \eta_s \eta_i \eta_r. \tag{2.27}$$

The three efficiency components of η_0 consider carrier loss due to lateral current spreading (η_s), barrier recombination and other leakage currents (η_i) and non-radiative recombination in the QW (η_r) [74]. The presented increased lateral SHB due to near-field narrowing at high currents can thus be incor-

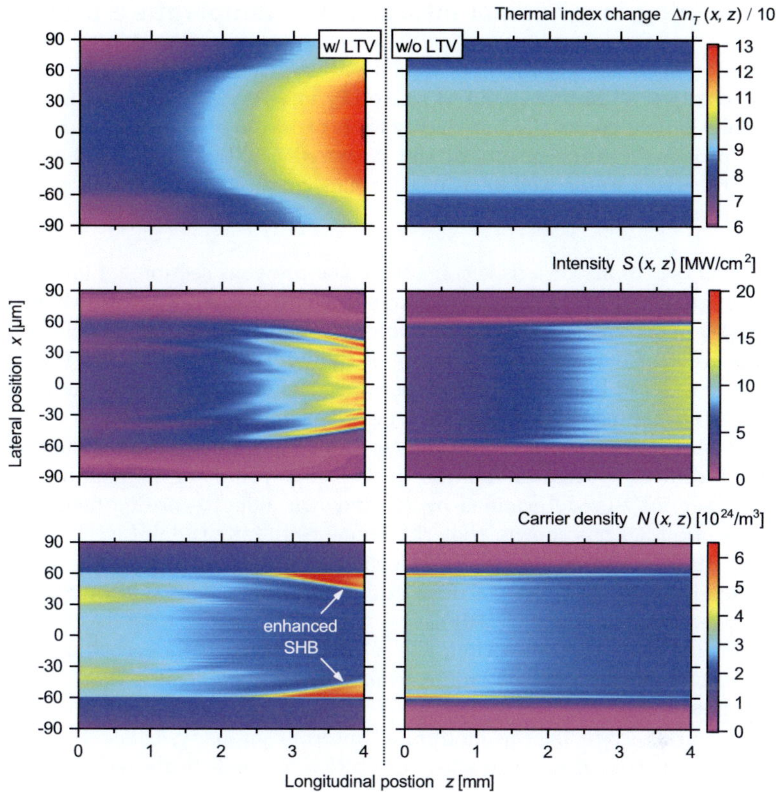

Fig. 2.11.: Effective refractive index change (top), internal intensity (center) and carrier density (bottom) for $I = 15.8\,\mathrm{A}$ simulated with the extended BALaser software. Published in excerpts in [75].

porated in the factor η_r which hence becomes current-dependent. However, it is hardly possible to experimentally separate and quantify such a decrease from the dominating effects limiting optical power increase, namely longitudinal SHB [89], thermal gain reduction [86], and waveguide recombination [94]. Because at least two of them coexist by nature. From the simulation data, however, the reduction of η_r by lateral SHB can be approximated by the share of the sindicated red-colored area in the Fig. 2.11 (bottom, left-hand side) over the total contact area.

2.5. Tailoring longitudinal intensity and temperature profile

This section deals with tailoring of the lateral-longitudinal temperature distribution to target an improvement of reliability and beam quality of EC-BALs. The analysis focuses on the homogenization of internal intensity and carrier density profiles by adjustments of the facet reflectivities for a fixed vertical and lateral structure as well as a fixed cooling architecture of the laser as introduced in section 2.3.

As far as the resonator axis is concerned, the previous section 2.4 has shown that a longitudinal isothermality has an even more valuable feature than just leading to a more efficient heat extraction and thus to an overall cooler device: It can prevent a thermal down-tapering of the lateral waveguide and thus improve the expected lifetime. The key parameter that trigger this phenomenon is the internal intensity S: Either by an explicit dependence or by an implicit influence on the carrier density by gain saturation, $S(z)$ determines the distribution of all heat sources in equation (2.18). The profile of internal intensity itself is essentially predetermined by R_b and R_eff [66]. Regarding beam quality, the section 2.4 has shown, that the influence of longitudinal isothermality itself on $\mathrm{BPP}_\mathrm{lat}$ is small. Since the following analysis tailors $T(z)$ via $S(z)$, it will hence be possible to gauge if an additional homogenization of $S(z)$ and $N(z)$ is beneficial in terms of $\mathrm{BPP}_\mathrm{lat}$.

Concerning the lateral direction, $T(x, z)$ develops basically according to the cooling geometry, the corresponding heat transfer coefficient $\kappa(x, y)$ and the dependence of the heat power density H on x. Inside the contact stripe, the latter is characterized by the modulated internal intensity, called filamentation, which for the thermal analysis of section 2.4 was neglected by spatial averaging. Such filaments are local intensity peaks which can be attributed to the nonlinear respective interaction of gain, refractive index and intensity in combination with a multi-modal lateral waveguide. Filamentation and thus the optical load can be enhanced by optical feedback [62,67] compared to free-running operation. Therefore, the elevation of the mean lateral intensity by an actual three-mirror cavity configuration is finally investigated by means of experimental near-field profiles.

2.5.1. Symmetric vs. asymmetric facet reflectivities

In this subsection, two types of coating layouts are compared which are referred to as symmetric and asymmetric in the following. The asymmetric design is that of section 2.4.2 with a back-side reflectivity close to unity ($R_\mathrm{b} = 96\,\%$) and a front-facet reflectivity $R_\mathrm{f} = 2\,\%$. It is commonly used to couple out the whole laser light mainly through one facet. The symmetric design is charac-

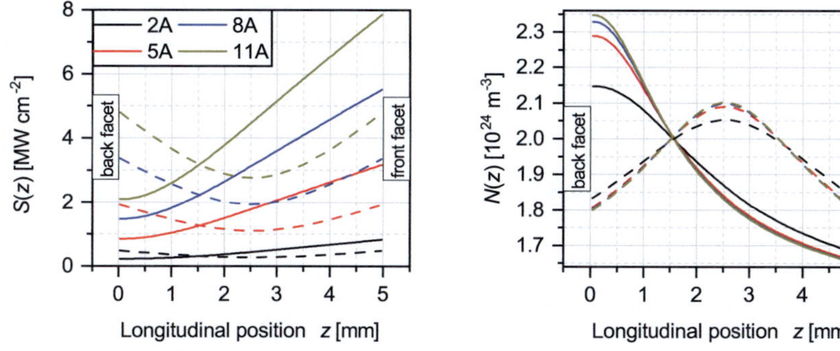

Fig. 2.12.: One-dimensional simulation of internal intensity (left) and carrier density (right) for symmetric ($R_{f,b} = 10\,\%$, dashed curves) and asymmetric ($R_f = 2\,\%$, $R_b = 96\,\%$, solid curves) coating design, neglecting thermal and lateral effects.

terized by nominal equal coatings $R_f = R_b$ which are chosen in a way that the corresponding mirror loss α_m is comparable to that of the asymmetric approach. As a result, equal power is coupled out from the two facets.

In the following experimental multi-device study, an symmetric design with $R_f = R_b = 10\,\%$ is compared to the asymmetric design to verify experimentally the simulative predictions from subsection 2.4.3. For this comparison, gain-guided DUTs with $L_i = 5\,\text{mm}$ and $w_c = 140\,\mu\text{m}$ are studied at a heat sink temperature of $25\,°\text{C}$. Due to manufacturing convenience the submount lengths are $5.4\,\text{mm}$ and $5\,\text{mm}$ for the asymmetric and symmetric design, respectively. To ensure comparability, it has been verified exemplarily that the influence of shortening the submount of the asymmetric devices to a length of $5\,\text{mm}$ is negligible with respect to their emission properties.

Analysis of optical power performance

For these device parameters, Fig. 2.12 illustrates the profile of $S(z)$ and $N(z)$ for an exemplary set of injection currents. These numerical computations are based on a simplified one-dimensional model [88] neglecting thermal and lateral effects. Details regarding this model and the numerical parameters used for the calculations can be found in the appendix A (Tab A.2). From Fig. 2.12 it is obvious that for the symmetric resonator the differences between maximum and minimum value of $S(z)$ and $N(z)$ are by about two times smaller, i.e. they are distributed more homogeneously compared to the asymmetric cavity.

The experimental data obtained for this comparison is shown in Fig. 2.13 as a function of injection current. The depicted curves are obtained averaging the data from at least four devices each. The error bars shown throughout this thesis represent one standard deviation in each direction. For the sake of visibility only a selected set of error bars are depicted.

The Figs. 2.13a and b show the basic power-related characteristics P_opt, U_0, η_eo as well as the change in T_jun with reference to T_hs and the corresponding

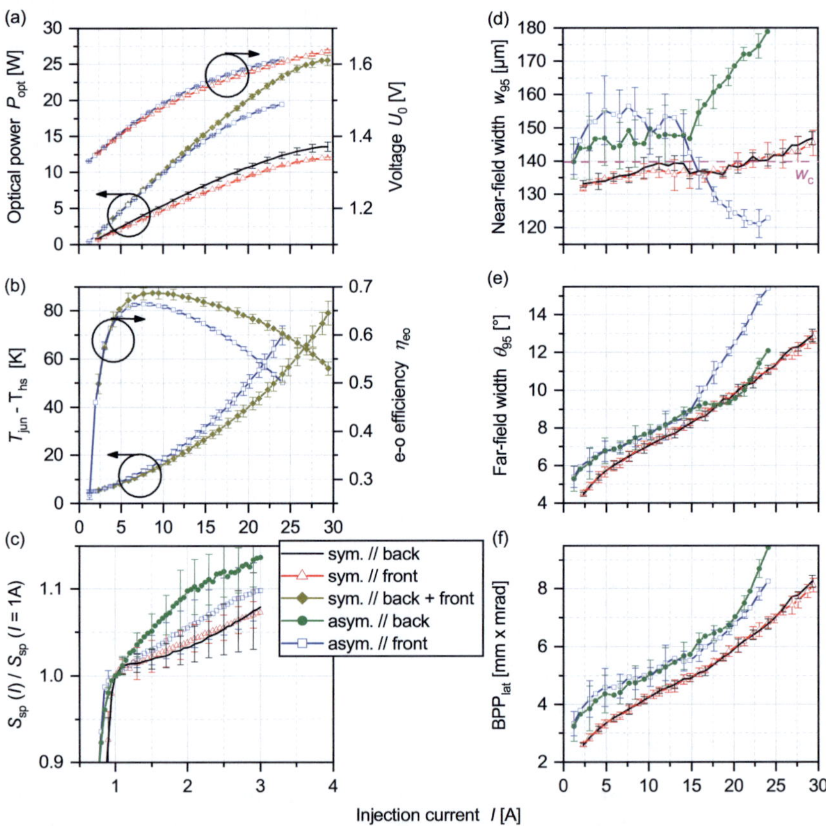

Fig. 2.13.: Experimental comparison of gain-guided BALs in free-running operation with asymmetric ($R_\mathrm{f} = 2\,\%$, $R_\mathrm{b} = 96\,\%$) and symmetric ($R_\mathrm{f,b} = 10\,\%$) coating designs for $T_\mathrm{hs} = 25\,^\circ\mathrm{C}$. The presented curves represent the average data from at least four devices each, [95].

basic characterization parameters are summarized in Tab. 2.1: The compared designs deviate by $15\,\%$ in terms of mirror loss α_m what results in an equal difference regarding I_th. The accompanying difference in external differential

Tab. 2.1.: Selected characteristics of gain-guided DUTs with asymmetric and symmetric facets coatings for $w_c = 140\,\mu\mathrm{m}$ and $L_i = 5\,\mathrm{mm}$. The comparatively high tolerances given for the R_s values are due to the nonlinearity of the voltage-current characteristics.

parameter	unit	asymmetric	symmetric
R_f		0.02	0.1
R_b		0.96	0.1
α_m	cm^{-1}	3.95	4.6
η_{ext}		0.85 ± 0.01	0.88 ± 0.01
I_{th}	A	0.85 ± 0.02	1.00 ± 0.02
R_s	$\mathrm{m\Omega}$	13 ± 6	12 ± 6
R_{th}	KW^{-1}	3.6 ± 0.2	3.5 ± 0.2

efficiency $\eta_{\mathrm{ext}} = U_{\mathrm{ph}}^{-1}\partial P_{\mathrm{opt}}/\partial I$, however, is negligible: The L-I curves basically overlap till $\sim 10 I_{\mathrm{th}}$ indicating a good comparability of these devices. Here, the numerical values for η_{ext} are obtained from a linear fit to the L-I curves till 5 A.

For the asymmetric design, the thermal roll-over of the optical power appears at lower I but approximately the same T_{jun} compared to the symmetric DUTs due to increased longitudinal SHB losses. The improved L-I curve is the main reason for an increasing difference in the e-o efficiency for the two approaches towards high currents, as can be seen in Fig. 2.13b. These power and efficiency benefits of the symmetric design are consistent with a previous study [88]. The novelty of the present study is the experimental analysis of the influence of facet symmetrization on the lateral beam parameters, presented in the following.

Analysis of near-field characteristics

In the Figs. 2.13d–f the near-field, far-field and $\mathrm{BPP}_{\mathrm{lat}}$ characteristics at both output facets are compared for both coating designs as a function of I. The near- and far-field widths of the asymmetric design develops just as shown in Fig. 2.8a with the divergence of front- and back-side widths starting around 15 A. In contrast, for the symmetric design, the values for w_{95} and θ_{95} at both facets basically overlap, developing linearly with current. Here, the maximum deviation of w_{95} and w_c is only 10 % for the whole investigated current range.

To connect these observations to the corresponding temperature distributions, Fig. 2.14 shows for both designs the simulation of $T(x, z)$ for a current of 20 A. There, it is evident that for the symmetric case, the longitudinal distribution is nearly symmetric with reference to the cavity center and is much more homogeneous compared to the asymmetric case. Hence, the symmetric

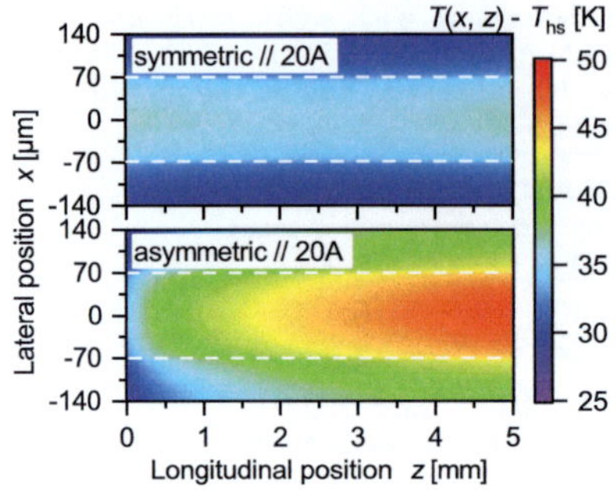

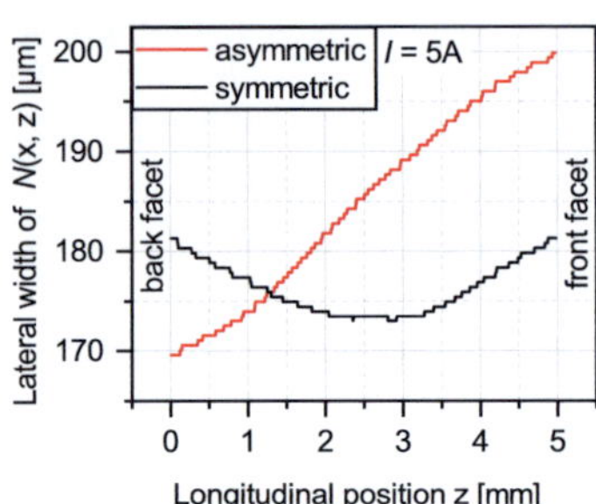

Fig. 2.14.: Temperature distributions for asymmetric and symmetric cavities at $I = 20\,\mathrm{A}$, simulated with the SEMSIS package. The edges of the contact stripe are marked with white dashed lines. [87]

Fig. 2.15.: Width of the lateral carrier density profile (95 % p.c.) at current of 5 A, simulated with the SEMSIS package.

case is a good approximation of the situation without LTV as simulated in subsection 2.4.3.

For $I < 15\,\mathrm{A}$, the values for w_{95} at the front facet of the asymmetric cavity are 5 to $20\,\mu\mathrm{m}$ larger than in the symmetric case and additionally exceed the width of the contact stripe. This observation indicates a larger lateral extension of the $S(x, z)$ throughout the asymmetric cavity, which is investigated in detail in the following, both via simulations and in the experiment.

Figure 2.15 depicts the simulated widths of $N(x, z)$ along the resonator for both coating designs at $I = 5\,\mathrm{A}$, including 95 % carrier content. These calculations are carried out with the SEMSIS package and neglect lateral current spreading. For both designs, the curves basically follow the shape for the corresponding internal intensity in the sense that a broad $N(x, z)$ distribution goes with a high intensity and vice versa (cf. Fig. 2.12).

In general, different widths of $N(x, z)$ exceeding w_{c} indicate a different amount of lateral carrier accumulation beside the contact stripe. Essentially, there are three mechanisms to produce such electron-hole pairs in the QW beside the contact: Carrier diffusion in the QW, electrical pumping via lateral current spreading and optical pumping due to high-divergence contributions of the electric fields not confined by the effective-index waveguide.

Since, for both designs, the carrier diffusivity D_N is chosen equal and current spreading is neglected in the simulation and is considered to be very similar in experiment, the differences in the widths of $N(x, z)$ illustrated in Fig. 2.15 can be attributed to optical pumping. What is more, the amount of optical power outside the contact scales strictly monotonically with optical power inside the contact opening, which becomes obvious when the Figs. 2.12 and 2.15 are compared. As a result, for gain-guided BALs a more homogeneous distribution of $S(z)$ leads to a narrower near field at the laser front in the current range where thermal lensing is weak and does not suppress lateral optical pumping of passive sections. This result is also confirmed experimentally by a measurement of the integrated rates of spontaneous emission S_{sp}, illustrated in Fig. 2.13c:

Since both coating designs have the same epitaxial, lateral and longitudinal structure they exhibit approximately the same amount of lateral current spreading beside the contact opening. Thus the relative differences of the slopes $\partial S_{\mathrm{sp}}/\partial I$ for the the two designs can be attributed to an increase of the total carrier number beneath the contact stripe due to SHB or beside the stripe due to optical pumping. As illustrated in Fig. 2.12, for both designs the carrier density at the front facets changes hardly with increasing current due to gain saturation. Therefore, the larger slope of S_{sp} in Fig. 2.13c at the front facet of the asymmetric design confirms the simulation results of Fig 2.15.

The even increased difference of the slopes of S_{sp} at the back facets is an result of increased longitudinal SHB for the asymmetric design which is strongest at the back facet (cf. Fig. 2.12).

In sum, it should be emphasized that the wider near-field width of the asymmetric as compared to the symmetric cavity design in the low power range is due to optical pumping instead of lateral current spreading.

The analysis of S_{sp} can further be extended to the current region $>15\,\mathrm{A}$ where strong thermal lensing confines the light essentially within the bound of the contact stripe and suppresses optical pumping of the area besides. Consequently, following the same line of argument as above, in this current region, a change in S_{sp} that is different for the symmetric and asymmetric case can be attributed to a change of carrier number underneath the contact stripe.

Figure 2.16 shows the near-field width at the front facet and the corresponding value of S_{sp} for one exemplary DUT with symmetric and asymmetric cavity each. The values for S_{sp} are normalized to a current of approximately $2I_{\mathrm{th}} \approx 2\,\mathrm{A}$. On the shown scale, normalized S_{sp} values for both devices increase and basically overlap till the intersection of the corresponding near-field widths at $\sim 17\,\mathrm{A}$. From the intersection point towards higher currents, the growth of the S_{sp} values for the asymmetric device develops with a two-times larger slope than for $I < 17\,\mathrm{A}$. This change of growth rate indicates an increased overall carrier number within the contact stripe at the front facet – assuming no change for lateral current spreading. The correlated appearance of such an increase of N underneath the contact area and w_{95} narrowing beneath w_{c} confirms experimentally the simulated appearance of increased lateral SHB due to near-field narrowing (cf. red-colored area in plot of $N(x, z)$ simulated with LTV, Fig. 2.11).

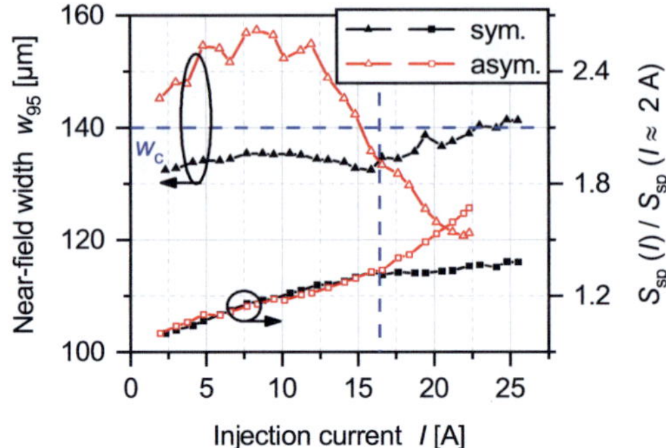

Fig. 2.16.: Experimental front-facet near-field width and corresponding normalized spontaneous recombination rate, compared for a single symmetric and asymmetric DUT each.

Analysis of far-field and BPP$_{\mathrm{lat}}$ characteristics

For the analysis of the near-field widths in Fig. 2.13 it has already been concluded that the longitudinal homogenization of the intensity and temperature profile by symmetric coatings is sufficient to show the characteristics as simulated without LTV in subsection 2.4.3.

The comparison of the far-field widths of the asymmetric and symmetric cavities, depicted in Fig. 2.13e, confirms this conclusion once more: The far-field blooming accompaning the near-field narrowing which is visible for the asymmetric design at the front facet at high currents, does not appear for the asymmetric DUTs. Here, as simulated for the isothermal case in Fig. 2.8b, the curves for θ_{95} at both facets overlap and increase nearly linearly with rising current.

However, in the current range $I < 15\,\mathrm{A}$ there is an offset between the far-field widths for both coating designs which cannot be caused by different longitudinal temperature distributions, as the simulations with and without LTV do not exhibit such an offset. Instead they develop at the same level with rising current before the blooming starts (cf. Fig. 2.8b). The cause of this offset between the far-field widths of the different coating designs is investigated in the following in more detail.

The fact that this offset is present also for the smallest currents studied indicates that it is of non-thermal origin. To separate thermal and non-thermal contributions to far-field blooming, the measurements of the lateral divergence of Fig. 2.13e are repeated in short pulse mode with a low duty cycle. Furthermore, the device-averaged beam parameters measured for ion-implanted DUTs with the asymmetric coating design were added to the previous comparison of gain-guided symmetric and asymmetric DUTs. The used type of implantation is known to suppress lateral carrier accumulation and therefore is expected to behave similarly to the symmetric cavity concerning θ_{95} in the absence of thermal effects.

To begin with, the rise of the junction temperature in short-pulse mode is investigated in order to assure that the employed pulse parameters enable a proper separation of thermal and non-thermal contributions to far-field blooming.

For the used pulse parameters (pulse duration $\tau = 4\,\mu s$, duty cycle $0.02\,\%$), the optical spectrum recorded with a spectrometer over an exposure time of several milliseconds will be a superposition of the instantaneous spectra of the laser. Since these spectra shift and scale in amplitude as the laser heats up during the current pulse, the centroid position of the time-averaged spectrum can only serve as an estimate for the mean T_{jun} during the pulse. Yet, if this cumulated spectrum can be temporally resolved, the maximum of $T_{\mathrm{jun}}(t)$ can be derived from the maximum of the centroid positions $\lambda_0(t)$ of the instantaneous spectra via equation (2.6).

In the experiment, spectral resolution of the optical pulses is achieved by passing a narrow-band monochromator ($0.01\,nm$ full width at half maximum (FWHM)) before being recorded by a photodiode. The transmission window of the monochromator is then successively tuned along the full spectral range of the DUT stimulated emission by 0.5-nm steps so that for each spectral position of the monochromator a time trace of the pulse is recorded. The result is illustrated exemplarily for an asymmetric DUT in Fig. 2.17.

The centered graph at the top shows the evolution of the emitted stimulated emission spectra along the optical pulse (bottom, center), normalized to its maximum for each temporal position t. The injection current is exemplarily set to $25\,A$. It is evident that λ_0 shifts to higher wavelengths till the pulse starts to fall at $t = 4\,\mu s$. The corresponding spectrum (top, left-hand side) is obtained by integration of the non-normalized data of the spectrally resolved pulse. Moreover, it is obvious that the maximum of the full spectrum is at about the same spectral position where the pulse reaches maximum intensity.

To estimate the maximum T_{jun} during a pulse, the peak centroid position within all the spectra of a pulse is evaluated for several currents, converted into temperature via equation (2.6) and plotted in Fig. 2.17 on the top right-hand side. There, the rise of T_{jun} with reference to T_{hs} is $6\,K$ at $20\,A$, which is about a ninth of the average CW value shown in Fig. 2.13b for the same current. If the graph of the top right-hand side of Fig. 2.17 is linearly extrapolated, it can be concluded that the rise of T_{jun} stays below $10\,K$ for $I < 30\,A$ and the given pulse parameters.

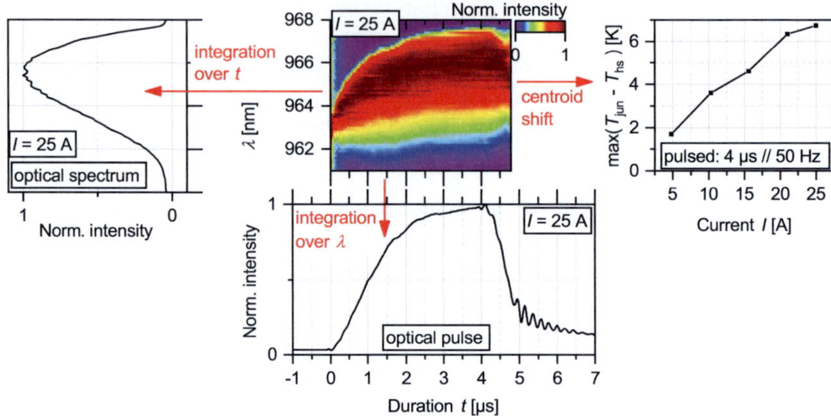

Fig. 2.17.: Spectrally resolved optical pulse for an exemplary asymmetric device ($w_\mathrm{c} = 140\,\mu\mathrm{m}$, $L_\mathrm{i} = 5\,\mathrm{mm}$) operating in short-pulse mode (25 A, $\tau = 4\,\mu\mathrm{s}$ at 0.02 % d.c.) (top, center): For each temporal position t the spectrum is normalized to its maximum value in order to reveal the spectral shift of the emission due to self-heating during the pulse. The corresponding full optical spectrum (top, left-hand side) and the full pulse shape (bottom, center) are obtained by integrating the non-normalized version of the spectrally resolved pulse over time t and wavelength λ, respectively. The maximum change of the junction temperature as a function of current (top, right-hand side) is extracted from the maximum shift of the centroid wavelength during a pulse.

In Fig. 2.18 P_opt and T_jun in CW as well w_{95}, θ_{95} and $\mathrm{BPP}_\mathrm{lat}$ for CW and pulsed operation are compared for ion-implanted asymmetric DUTs and gain-guided asymmetric and symmetric devices. The CW values for the latter group of devices are the same as in Fig. 2.13. The curves depicted in Fig. 2.18 represent a mean value obtained from at least four devices of each configuration. The comparison of the L-I curves of the three configurations in Fig. 2.18 reveals that for the ion-implanted version thermally induced reduction of the power slope appears at smaller currents than for the gain-guided structures. Since the thermal resistances of gain-guided and ion-implanted designs are $(3.6 \pm 0.2)\,\mathrm{K\,W^{-1}}$ and $(3.5 \pm 0.2)\,\mathrm{K\,W^{-1}}$, respectively, and thus are nearly equal, the reduced power slope is attributed to modifications of the QW by the employed proton bombardment [73].

For $I < 15\,\mathrm{A}$, the values of w_{95}, θ_{95} and $\mathrm{BPP}_\mathrm{lat}$ measured in CW and pulsed mode as a function of current develop at a lower level for the ion-implanted design but at a very similar shape as compared to the gain-guided DUTs

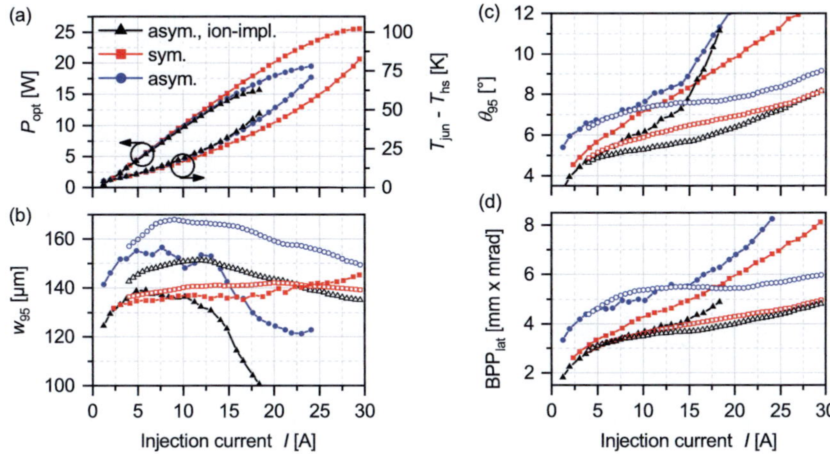

Fig. 2.18.: Comparison between ion-implanted asymmetric as well as gain-guided asymmetric and symmetric DUTs in CW (full symbols) and pulsed mode (open symbols) with a nominal pulse period of 4 µs and repetition rate of 50 Hz. The contact width is 140 µm and the cavity length 5 mm. The data are average values of at least four devices for each configuration. Error bars are of similar size as those shown in Fig. 2.13, but are left out here for the sake of visibility.

with asymmetric facet coatings. In CW mode, a similar reduction of θ_{95} and $\mathrm{BPP_{lat}}$ have been measured for deeply implanted BALs in [22] and have been attributed to a reduction of lateral carrier accumulation.

Apart from that, it is visible in Fig. 2.18 that in CW mode the slopes quantifying the near-field narrowing $\left|\frac{\partial w_{95}}{\partial I}\right|$ and accompanied far-field blooming $\left|\frac{\partial \theta_{95}}{\partial I}\right|$ are larger for the ion-implanted DUTs.

Concerning the far-field divergence in pulsed operation, it is observed that the θ_{95} values for the ion-implanted and the symmetric designs are almost equal at low currents ($\sim$5 A). As shown in Fig. 2.17, in this current range thermal effects are negligible ($T_{\mathrm{jun}} - T_{\mathrm{hs}} < 2\,\mathrm{K}$) so that the lateral far-field distribution is essentially carrier-related. Thus the equivalence of θ_{95} for the ion-implanted and symmetric design in the absence of thermal effects indicates that the mitigation of lateral carrier accumulation is similar for both structures. Consequently, the lower far-field widths measured for symmetric as compared to the gain-guided asymmetric design (cf. Figs. 2.13e and 2.18c) can be attributed to a reduction of lateral carrier accumulation by reduced optical pumping of the region adjacent to the contact stripe which, in turn, is a result of the longitudinal intensity homogenization by equal facet reflectivities.

Eventually, the lower near- and far-field widths of the symmetric approach lead to a $1\,\text{mm} \times \text{mrad}$ lower BPP_{lat} as compared to the gain-guided DUTs with asymmetric coatings: In the region $<15\,\text{A}$, before a narrowing of the near field below w_{c} occurs, this beam quality improvement is due to longitudinal intensity homogenization. For the high-current region $>20\,\text{A}$, on-going far-field blooming overcompensates the saturating near-field narrowing and thus leads to an increasing difference of the front-facet BPP_{lat} of the asymmetric and symmetric designs (cf. Figs. 2.13f and 2.18d). The latter effect occurs as predicted by the analysis of the highest longitudinal temperature gradient in Fig. 2.10.

An interesting side effect is visible in the analysis of the BPP_{lat} in Fig. 2.18d: Up to current values of $15\,\text{A}$ – as they are common for BALs of such stripe width on laser bars – the BPP_{lat} is primarily determined by the non-thermal and thus carrier-induced effects for the employed gain-guided asymmetric design.

2.5.2. Optical feedback by an actual coupled-cavity setup

So far, free-running BALs and EC-BALs have been treated likewise within the approximation of matched optical feedback including the compound-cavity model. However, it has to be experimentally verified whether these approximations can be maintained for actual coupled-cavity laser systems.

The advantages revealed for symmetric BALs in terms of beam quality and reliability rely on longitudinal intensity and temperature balancing and have been described within the model of matched optical feedback (cf. section 2.1). This framework allows to neglect coupling losses ($\Gamma_{\text{c}} = 1$) due to misalignment so that the investigated symmetry of $T(z)$, $S(z)$ and $N(z)$ is ensured for all injection currents. Actual external-cavity configurations (such as in Fig. 1.1), however, are by nature susceptible to thermo-mechanical disturbances that may lower Γ_{c}:

Thermal expansion due to self-heating of the laser and/or heating of cavity elements by absorption of laser light may cause relative translations and tilts of cavity elements and thus change the mutual mechanical arrangement and the beam propagation. As a consequence, the feedback light may hit the front facet with a spatial displacement which may cause local heating of adjacent layers and hence distort $T(x, z)$. Furthermore, a reduction of Γ_{c} can be treated as a reduction of R_{eff} which changes the longitudinal distribution of $S(z)$ and thus the heat power densities. Therefore it may be that the benefits found for symmetric double-sided BALs in case of matched optical feedback can hardly be transferred to EC-BAL devices in a realistic laboratory environment.

This is the reason why in the following, symmetric, coupled-cavity EC-BALs are experimentally compared to free-running DUTs with an equivalent mirror loss. In a second step, the lateral intensity filamentation, which is neglected in the steady-state compound-cavity model, is experimentally studied for free-running and external-cavity operation. The external-cavity configuration investigated in this chapter is shown in Fig. 2.2 and consists of a DUT, collimation lens L_1 and the external reflector E_1 with a reflectivity R_e. Analogously to the free-running case studied in subsection 2.5.1, EC-BALs having a longitudinally (a)symmetric internal intensity are referred to as (a)symmetric EC-BALs. The corresponding reflectivity designs (cf. Fig. 2.1) read $R_b \approx R_e$, $R_f \ll 1\%$ for the symmetric and $R_b > 95\%$, $R_f \ll 1\%$, $R_e = 1$ to 10% for the asymmetric configuration. In the following, these designs are benchmarked briefly concerning their suitability for wavelength stabilization.

Wavelength stabilization with symmetric EC-BALs

For many applications of wavelength-stabilized EC-BALs such as zero-phonon line pumping or DWBC it is essential that lasing around the stabilized wavelength λ_s can be provided for a large mismatch of λ_s and the gain peak wavelength λ_p to enable laser functionality at various levels of drive current and hence output powers [96]: The temperature dependence of λ_p requires that for wavelength-stabilized EC-BALs, spectrally matched feedback ($\lambda_p \approx \lambda_s$) can only be maintained for a restricted set of injections currents. The locking range T_{lock} determines how far λ_p may deviate from a fixed stabilized wavelength λ_s on condition that the lasing threshold for λ_s is lower than for λ_p. Thereby T_{lock} determines the accessible range of output powers for the condition $\lambda_0 \approx \lambda_s$.

Assuming constant spectral profiles of $R_{f,b}$ and a parabolic semiconductor gain curve with constant linewidth in the vicinity of λ_p, T_{lock} depends strictly monotonically on the change of g_{mod} by optical feedback compared to the free-running case, $g_{\text{mod,free-lock}}$, given by

$$\Delta g_{\text{mod,free-lock}} = g_{\text{mod}}(R_e = 0) - g_{\text{mod}}(R_e) =$$

$$= \frac{1}{2L_i} \ln\left(\frac{R_{\text{eff}}}{R_f}\right) \overset{R_f \ll R_e}{\approx} \frac{1}{2L_i} \ln\left(\frac{R_e}{R_f}\right), \tag{2.28}$$

which is derived from equation (2.4) for a spatially homogeneous g_{mod} as well as a constant internal absorption. Here, the AR coating of the front facet can be estimated by $R_f \leq 0.5\%$, which can be achieved with common three-layer coatings over a spectral of close to $100\,\text{nm}$.

Now, the advantage of double-sided symmetric versus single-sided EC-BALs becomes obvious from equation (2.28): For symmetric out-coupling, the mirror loss α_m is distributed over both facets so that a higher R_e is needed for the symmetric configuration to obtain the same α_m as the single-sided

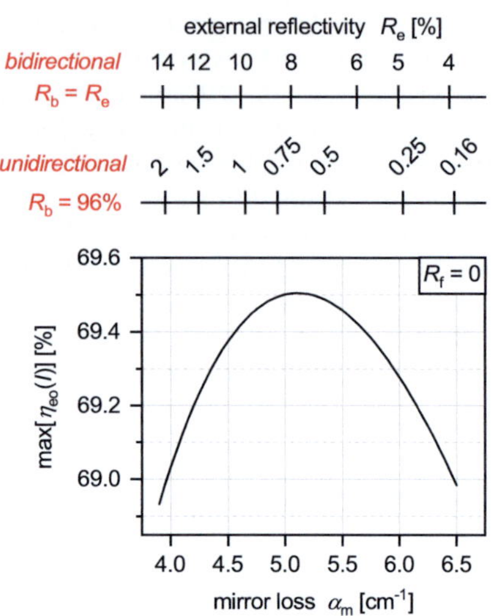

Fig. 2.19.: Calculated maximum electro-optic efficiency of single- and double-sided emitting EC-BALs as a function of mirror loss. The computations are based on lumped laser model summarized in the appendix A, Tab. A.3.

counterpart. As a result $\Delta g_{\mathrm{mod,free\text{-}lock}}$ and thus T_{lock} are higher for symmetric EC-BALs.

In case the same value of R_e is chosen for both EC-BALs designs, the unidirectional approach experiences more absorption losses as indicated by Fig. 2.19. This graph shows the maximum e-o efficiency as a function of α_m based on a simple lumped laser model. For details of this model and the numerical parameters used for the simulation see the appendix A, Tab. A.3. In Fig. 2.19 the corresponding values of R_e are additionally shown for asymmetric and symmetric EC-BALs. It is evident that for symmetric EC-BALs R_e can be chosen in the range from 4 to 14 % with the e-o efficiency being >69 %. Whereas for the single-side emitting configuration this efficiency limit can be maintained only for $R_f \leq 2\%$. This result reveals that such bidirectional EC-BALs provide a higher flexibility in terms of external reflectivity R_e, which e.g. can be used to pre-compensate a potential cavity misalignment with limited loss of e-o efficiency.

Im sum, as well in terms of the locking range EC-BALs profit from a symmetric allocation of the mirror loss to two out-coupling elements.

Tab. 2.2.: Selected characteristics of double-side out-coupling BALs in free-running and external-cavity configuration. The geometrical dimension of the DUTs are $w_\mathrm{c} = 130\,\mu\mathrm{m}$ and $L_\mathrm{i} = 5\,\mathrm{mm}$. In case of external-cavity operation, feedback is provided through the front facet and the external cavity length amounts to two times the focal length f_1 of the collimation lens, i.e. $w_\mathrm{e} = 2f_1 \approx 16\,\mathrm{mm}$. The actual values of the facet reflectivities, measured with ellipsometry, are put behind the target values in brackets.

parameter	unit	free-running gain-chip	symmetric external cavity	symmetric free-running
R_b	%	4.5 (4.1)	4.5 (4.1)	4.5 (3.7)
R_f	%	0.5 (0.5)	0.5 (0.5)	4.5 (5.4)
R_e	%	–	5.2	–
α_m	cm^{-1}	8.4 (8.5)	6.0 (6.1)	6.2 (6.2)
I_th	A	1.40 ± 0.05	1.14 ± 0.05	1.15 ± 0.05
R_s	mΩ	13 ± 8	14 ± 9	14 ± 9
R_th	KW^{-1}	4.1 ± 0.2	4.1 ± 0.2	4.1 ± 0.2
		front back	front back	front back
η_ext		0.66 0.24	0.45 0.49	0.46 0.49

2.5.2.1. Symmetrizing longitudinal intensity profile by optical feedback

Symmetric EC-BALs are studied experimentally in the configuration depicted in Fig. 2.2 with the external reflector E_1 inserted. In this case E_1 is a broadband dielectric mirror with a reflectivity of $R_\mathrm{e} = 5.2\,\%$. The semiconductor chips are laterally gain-guided with the following characteristics: $L_\mathrm{i} = 5\,\mathrm{mm}$, $w_\mathrm{c} = 130\,\mu\mathrm{m}$, $R_\mathrm{b} = 4.5\,\%$, $R_\mathrm{f} = 0.5\,\%$. For $T_\mathrm{hs} = 25\,^\circ\mathrm{C}$, these devices are characterized in free-running and external-cavity configuration concerning P_opt, U_0, e-o efficiency, T_jun as well as w_{95}, θ_{95} and $\mathrm{BPP}_\mathrm{lat}$ and are compared to symmetric free-running DUTs with $R_\mathrm{f,b} = 4.5\,\%$.

The results of this study are shown in Fig. 2.20 as a function of injection current. In Fig. 2.20a the optical output power extracted from back and front facets is shown. It can be seen that inserting the feedback mirror E_1 symmetrizes the power extraction from both facets. The corresponding external differential efficiencies η_ext, extracted from a current interval from 2 to 5 A, are summarized among other basic characterization data in Tab. 2.2. By means of 5.2 % optical feedback, the back-facet η_ext could be raised to about the same level as in case of the symmetric free-running DUTs. The fact that, for the latter devices, the values of η_ext at both facets are not exactly equal is due to process-related tolerances of the coating reflectivities so that the actual values deviate slightly from the target reflectivities.

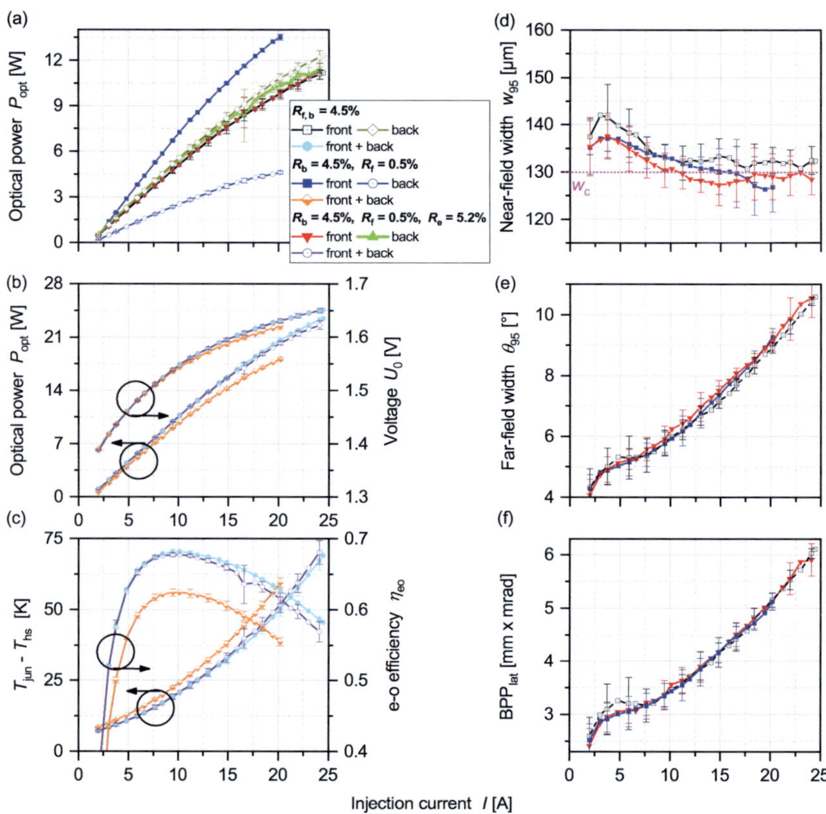

Fig. 2.20.: Experimental comparison of double-sided out-coupling BALs for $T_\mathrm{hs} = 25\,°\mathrm{C}$: Free-running DUTs with symmetric facet coatings ($R_\mathrm{f,b} = 4.5\,\%$) are compared to DUTs in external cavity configuration where a symmetric power extraction is established by an external mirror with a reflectivity $R_\mathrm{e} = 5.2\,\%$ in conjunction with an AR-coated front-facet with $R_\mathrm{f} = 0.5\,\%$. As a reference, the latter DUTs are also characterized in free-running operation. The portrayed data are average values from four devices of each type. Basic characterization data are listed in Tab. 2.2.

Furthermore, in the Figs. 2.20b–c it can be seen that the symmetric EC-BALs have nearly equal characteristics in terms of P_{opt}, U_0, η_{eo} and T_{jun} as the symmetric DUTs in free-running mode. In contrast, the sole free-running gain chips run less efficiently and thus at a lower output power and higher junction temperature because their mirror loss is about one third larger. Here, it has to be noted that for current values $>15\,\text{A}$ a mechanical re-alignment of the external-cavity mirror E_1 had to be conducted at each current step to counteract thermo-mechanical misalignment and assure maximum Γ_c.

In the Figs. 2.20d-f it is illustrated that with regard to beam quality and the corresponding beam parameters w_{95} and θ_{95}, all three configurations behave equally as a function of injection current and show the same trends as the free-running symmetric devices with $R_{f,b} = 10\,\%$ at current levels $I > 15\,\text{A}$ (cf. Fig. 2.13). Particularly, there is no indication of near-field narrowing accompanied with additional far-field blooming in the high current range. These observations lead to the following important conclusions:

- The beneficial impact of longitudinal intensity symmetrization on beam quality and reliability studied in the framework of matched optical feedback can be transferred to EC-BALs in real laboratory environment without negative influence on beam quality if external-cavity misalignment can be restricted. Based on these results, bidirectional wavelength-stabilized EC-BALs are feasible that outperform conventional single-side EC-BALs in all parameters related to power, beam quality, reliability as well as locking range. Yet double-sided out-coupling requires additional optical elements to bundle both outputs which leads to potentially higher device expenses and lower brightness compared to unidirectional laser configurations. Here, polarization beam combining would be a suitable means to combine the two outputs [97].

- Even for the worst case of total feedback mismatch ($\Gamma_c = 0$) the coating asymmetry of the sole gain chips in Fig.2.20 is small enough so that near- and far-field widths as well as the BPP_{lat} remain nearly unchanged compared to the case of fully matched feedback. Potential re-absorption of misaligned feedback light is neglected in this study (cf. chapter 3).

- Due to the selection of a rather low external reflectivity $R_e \approx 5\,\%$ the maximum misaligned power in the worst case of $\Gamma_c = 0$ would be only about $75\,\%$ of that present with single-side emitting EC-BALs (given by the ratio of the front-side η_{ext} for the free-runnuing gain chip in Tab. 2.2 and that of the asymmetric devices studied in Tab. 2.1).

2.5.2.2. Lateral intensity filamentation

In general, the lateral non-uniformity of the near-field distribution exhibiting several peaks and valleys is referred to as filamentation.

On the picosecond level, lateral intensity modulation is caused by the nonlinear interaction of the electric field and gain. Here in particular, the dynamical interplay of field diffraction within the gain aperture, intensity induced gain saturation as well as carrier-induced refractive index changes has been identified as a cause for this complex spatio-temporal behavior [40–42, 67, 98, 99]. Moreover, on the time scale of several 100 ns [100], the field distribution is coupled to thermal refractive index changes via the various loss mechanisms stated in equation (2.18). Since the carrier-induced filamentation takes place on time scales that are at least two orders of magnitude slower than thermal effects, it is a valid approximation that the quasi-stationary field distributions, time-averaged over $\sim$1 ns, determine the profile of the heat sources and thus Δn_T (cf. subsection 2.4.1).

The long-term near field captured on the time scale of several milliseconds to seconds, in turn, represents an additional time-average over the thermal dynamics. However, the deviations of the heat sources densities are small, once the comparatively large volume of the laser fixture has sufficiently approached thermal equilibrium and provided that the internal laser parameters do not degrade.

As a consequence, the lateral intensity pattern as recorded with the experimental setup in Fig. 2.2 provides the relevant information to detect lateral peaks of optical and thermal load relevant for device degradation.

The obvious disadvantage of a filamented near-field profile in terms of reliability is that the optical and thus thermal load is locally higher than for a uniform distribution of same lateral extension and power content. Whereas, the software packages used in this thesis can reproduce the widths of experimental near-field profiles accurately, they are lacking in reproducing the exact shape of the experimental profile. Therefore, filamentation is investigated here with the help of experimental time-averaged near-field profiles measured for the symmetric DUTs characterized in the previous subsection 2.5.2.1.

In general, the lateral intensity modulation is fully characterized by its spatial frequency spectrum. For the prediction of the filamentory behavior as function of e.g. current, the full spectral information, however, is not necessary. Therefore theoretical papers use a sinusoidal description with a single spatial frequency and amplitude to model the intensity modulation which allows a pertubational treatment [99, 101]. Since, however, experimental modulation profiles significantly deviate from the simple sine-shape, experimental treatments have proceeded to quantify filamentation by cumulated parameters such as the standard deviation of the lateral average of $S(x)$ at the facet position [62] or a single-peak analysis [102].

In terms of reliability, the deviation of the lateral peak intensity S_{max} with reference to the average lateral intensity S_{avg} is most critical. Therefore, in this thesis, the filamentation strength is characterized by a parameter A_{fil} defined as

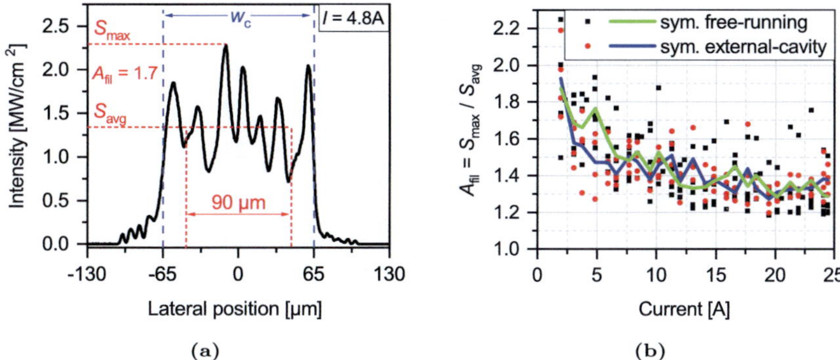

Fig. 2.21.: Quantification of experimental filamentation for symmetric DUTs with $L_\mathrm{i} = 5\,\mathrm{mm}$ and $w_\mathrm{c} = 130\,\mu\mathrm{m}$: Derivation of the parameters S_max and S_avg by which the filamentation strength A_fil is quantified (a) and experimental values for A_fil are given as a function of current (b). Solid lines represent average values from four devices for each configuration, whereas the dots depict data points from individual devices.

$$A_\mathrm{fil} := \frac{S_\mathrm{max}}{S_\mathrm{avg}}. \tag{2.29}$$

Here, S_avg is obtained from the central region of the near field that has a size of $40\,\mu\mathrm{m}$ less than the corresponding contact width to exclude a variation of the edge regions from affecting the averaging process. The employed evaluation scheme for quantification of A_fil is illustrated in Fig. 2.21a for a DUT with $w_\mathrm{c} = 130\,\mu\mathrm{m}$.

Figure 2.21b shows the filamentation strengths experimentally obtained for a sample set of the symmetric BALs in free-running and external-cavity operation characterized in Tab. 2.2. The corresponding device parameters are $L_\mathrm{i} = 5\,\mathrm{mm}$ and $w_\mathrm{c} = 130\,\mu\mathrm{m}$. It is evident that the average A_fil values for external-cavity and free-running operation nearly overlap, decrease with increasing pumping till $15\,\mathrm{A}$ and get saturated in the range of $A_\mathrm{fil} = 1.3$ to 1.4 towards higher current values. Moreover, the scattered individual data points show that optical feedback from a 2f external cavity does not change the filamentation strength significantly for the specific devices examined. In fact, the majority of the outliers to values higher than the average A_fil in Fig. 2.21b are caused by free-running devices. These observations agree with findings of a modal analysis [67] that matched optical feedback, indeed, changes the amplitudes of the excited set of modes slightly. Yet, strong excitation of a specific lateral mode that prevails over all other modes by multiple integer times, can only be achieved by asymmetric external-cavity configurations that effectively filter the feedback in spatial or angle space [45, 47, 103].

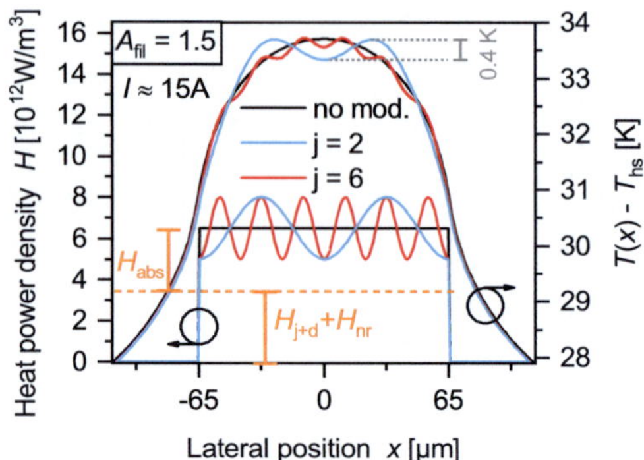

Fig. 2.22.: Impact of lateral intensity filamentation on lateral temperature profile, simulated with the thermal solver of the extended BALaser package for a free-running BAL with $w_{\mathrm{c}} = 130\,\mu\mathrm{m}$, $L_{\mathrm{i}} = 5\,\mathrm{mm}$ and $R_{\mathrm{b,f}} = 4.5\,\%$ operating at 15 A. A homogenous distribution of the heat power density along the contact stripe (black line) is compared to a $\sin^2$-shaped modulation of H_{abs} where the modulation period amounts to w_{c} over j. H is assumed constant along the longitudinal axis.

The results portrayed in Fig. 2.21b suggest that a value of $A_{\mathrm{fil}} = 1.5$ can be considered to be representative for an estimation of the impact of filamentation on $T(x)$, which is illustrated in Fig. 2.22. In this example, $T(x)$ is calculated for an exemplary operation current of 15 A for a symmetric BAL in free-running operation with $R_{\mathrm{f,b}} = 4.5\,\%$. Here, the thermal solver of the extended BALaser package is used where the heat transfer coefficient h_{hs} is chosen so that the simulated values for $\overline{T}_{\mathrm{QW}}$ match the T_{jun} shown in Fig. 2.20c. Furthermore, it is assumed that 55 % of the heat power belong to the sum of $H_{\mathrm{j+d}} + H_{\mathrm{nr}}$ which are assumed constant over the contact area. For the remaining 45 % of the heat power that make H_{abs}, a sum of an offset B and a $\sin^2$-shaped modulation with a amplitude A is chosen. Here, A and B are related via $A_{\mathrm{fil}} = (A + B)/(A/2 + B) = 1.5$.

The cumulated heat power density H is depicted in Fig. 2.22. There, the period of the modulation is set to w_{c} over j, where the cases $j = 2, 6$ are contrasted to a stepwise constant distribution of H_{abs} (black lines). It is obvious from the corresponding temperature profiles in Fig. 2.22 that the shorter the modulation period, the lower the deviations from the non-modulated $T(x)$ profile become. But even the highest modulation amplitude visible is only $0.4\,\mathrm{K}$ which is equivalent to only 1 % of the central temperature rise $T(x = 0) - T_{\mathrm{hs}}$. The reason for this comparatively small impact of filamentation on $T(x)$ is the smoothing property of the heat conduction equation (2.17).

From these observations, it can be concluded that filamentation as recorded for free-running and external-cavity operation in Fig. 2.21b can be neglected for the evaluation of the steady-state internal temperature profile and likewise for thermal refractive index changes. Apart from this, Fig. 2.22 shows that filamentation can displace the maximum QW temperature from the stripe center towards the position of maximum near-field intensity. Thereby lateral filamentation can influence the potential lateral onset position of COMD, as will be shown in section 3.1.

3. BALs subject to spatially mismatched optical feedback

In general, back-irradiance that re-enters the semiconductor waveguide only partly or not at all poses a serious threat to laser lifetime. For radiation incident on waveguide surroundings may cause strong local heating by highly-absorbing materials. Such localized heat sources elevate the temperature in the facet region so that a lower amount of laser self-heating of the bulk material is sufficient to trigger CO(M)D as compared to free-running operation.

Moreover, displaced back-irradiance can access and thus be strongly amplified by regions of unsaturated gain which in free-running operations originate at the edges of the contact opening. Thereby feedback to these regions can help higher-order lateral modes to reach threshold and hence possibly degrade beam quality.

The sources of such undesired back-irradiance are numerous and range from residual reflectivities of optical elements over back-scattering from a potential workpiece to intentional optical feedback misaligned by thermo-mechanical expansion of external-cavity elements.

In case of EC-BALs, the latter is the most critical source since feedback strengths up to 10 % of the total output power are common and the feedback return spot is highly focused. Especially for wavelength-stabilized BAL bars that employ compact 2f external-cavity configurations (cf. Fig. 1 in [104]), feedback misalignment is automatically accompanied by diode bar smile [34] which describes the effect that emitters on a bar may have different vertical positions due to mechanical strain.

There are several experimental studies confirming that optical feedback accelerates the degradation rate of semiconductor lasers, thus shortening the time to CO(M)D [15, 16, 34, 64]. For matched feedback, a theoretical study considers the enhanced internal intensity due to feedback to be responsible for an enhanced degradation rate [105]. However, the available experimental data suggests that misalignment makes a major contribution to laser degradation under feedback: By the analytic formulas provided in [66], it can be assessed that the increase of the internal intensity for single-side emitting BALs is $\leq 20\,\%$ for $R_\mathrm{f} = 0.5\,\%$ and $R_\mathrm{eff} \leq 10\,\%$. It has been shown that for state-of-the-art, 95 µm-wide and 3 mm-long BALs at 976 nm, comparable increases of the internal intensity by injection current during a step-test do not lead to significant device failures up to a current of 14 A after 15 kh in free-running operation at 50 °C heat-sink temperature [20]. Likewise, free-running, 100 µm-wide and 4 mm-long BALs from a different manufacturer have proven to operate without failure for 15 kh at 8 W output power and 20 °C. Though, when the latter BALs are subjected to 4 % feedback and are assembled on bars, they begin to show increasing emitter failure already after some hundred hours [16, 106].

However, the absorption of feedback light by the waveguide surroundings depends strongly on the employed combination of semiconductor and packaging material: For example, for GaAs-based laser diodes emitting at higher photon energies than the GaAs bandgap ($1.425\,\mathrm{eV} \mathrel{\hat=} 870\,\mathrm{nm}$), the substrate is opaque in terms of inter-band absorption with an corresponding absorption coefficient in the range of $10\,000\,\mathrm{cm}^{-1}$ [107]. In fact, for BALs emitting around 800 nm, it has been shown that COD is stimulated by detuning the feedback return spot towards the substrate [108]. By means of thermoreflectance measurements [109] it has been confirmed that the highest amount of device failures coincides with the largest rise of facet temperature. Thereby that study concludes that CO(M)D by feedback is thermally activated with activation energies that are in the same range as in free-running operation.

For the strained InGaAs/AlGaAs material system with emission wavelengths in the range of 900 nm to 1100 nm the situation is different. Here, all semiconductor layer are transparent for the laser emission in terms of interband transitions so that the radiation is primarily attenuated by the weaker process of intraband free-carrier absorption (cf. equation (2.9)). Furthermore, InGaAs-based laser diodes have proven to be in general more resistant to COD than AlGaAs-based lasers emitting in the range of 800 nm [33]. Consequently, the impact of feedback misalignment on the degradation behavior of InGaAs-based high-power laser diodes has remained an open question so far.

Therefore, the influence of spatial misalignment of the feedback return spot is investigated in more detail in the following for the InGaAs/AlGaAs-based DUTs introduced in section 2.3. The role of absorption by the different materials used for the employed laser package is analyzed by irradiating a DUT by a heating laser. The spot of the heating laser is swept vertically along the facet of a DUT. For each vertical position, non-destructive thermoreflectance and destructive damage-threshold measurements are conducted to identify weak spots.

Concerning a lateral deviation from the condition of matched feedback, most reports focus on filtering of the feedback beam in angle space to select a specific lateral mode and thus improve beam quality [45, 47, 50], or to control the spatio-temporal emission dynamics [103, 110]. However, a spatial displacement of the feedback return spot in the lateral direction has so far been reported only with respect to accompanying multi-path interference in the external cavity and the corresponding dynamics [111]. For high-power EC-BAL based on a laterally gain-guided semiconductor structure, however, a lateral feedback misalignment can modify the beam parameters. This is because a laterally displaced feedback beam may be amplified stronger compared to a aligned one due to lateral SHB and because the internal electrical field is only confined by a flexible thermally induced waveguide instead of a fixed built-in index waveguide. Hence for EC-BALs in high-brightness applications where the usable BAL optical power is limited by its BPP_{lat}, it is essential to know the change of the lateral beam parameters that is accompanied by an potential lateral displacement of back-irradiance.

Therefore, after vertical also lateral feedback displacement is studied in the following for the gain-guided DUTs.

3.1. Vertical displacement

In this section, the influence of a vertically displaced feedback return spot is investigated for the BAL structure introduced in section 2.3. For this investigation, gain-guided, single-side emitting DUTs emitting around 950 nm with $L_i = 4\,\text{mm}$, $w_c = 100\,\mu\text{m}$, $R_b = 96\,\%$ and $R_f = 2\,\%$ are selected and a heat sink temperature of $23\,^\circ\text{C}$ is chosen. The corresponding thermal resistance is $(4.5 \pm 0.2)\,\text{K W}^{-1}$ for p-side down mounted DUTs and about 2.3 times higher for the same device structure mounted p-side up.

3.1.1. Experimental setup

The experimental setup employed for this study is shown in Fig. 3.1. It is basically an extension of the setup shown in Fig. 2.2, where the imaging setup to determine the lateral beam parameters has been excluded from the drawing for illustration purposes. With this setup, the impact of a displaced laser beam on the facet of a DUT can be studied by means of optical feedback or optical injection by a second BAL referred to as heating laser in the following.

Optical feedback can be provided via the external cavity terminated by the external reflector E_2, which in this case is a broad-band HR mirror. Whereas the lateral axis is imaged telecentrically on E_2 by the cylindrical lens L_4 with a focal length $f_4 = 250\,\text{mm}$, the vertical axis is collimated by L_1, which is the same aspheric lens as introduced in section 2.2. By this lens configuration, the feedback spot on the DUT facet may be displaced by tilting E_2 around its lateral axis. The feedback strength, which is quantified by the effective reflectivity R_e of the external cavity, may be varied by use of a quarter-wave plate ($\lambda/4$) in combination with a PBS and is obtained via

$$R_e = R_{\text{BS}_1} \frac{P_2}{P_1}. \tag{3.1}$$

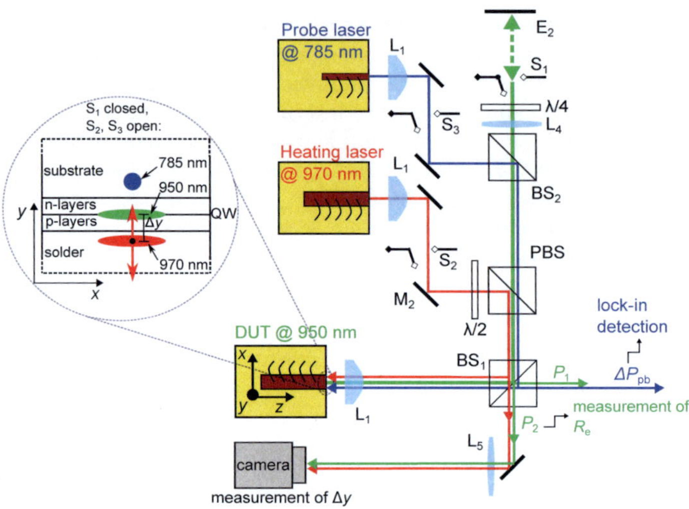

Fig. 3.1.: Schematic setup for the investigation of facet temperature subject to optical feedback and injection via thermoreflectance: By means of a probe laser (blue) that is incident on the DUT substrate, the change of front-facet reflectivity with temperature is transformed into a change of power ΔP_{pb}. Via the shutters $S_{1,2}$, the DUT front facet may be irradiated by a second laser source (red) or by optical feedback from an external cavity (green). Displacements Δy of the corresponding spots on the DUT facet are achieved via tilt of the mirrors M_2 and E_2, respectively, and can be monitored with a camera in the transmission path of BS_1 by a near-field imaging with a magnification of 62.5. Published in excerpts in [112].

Here, R_{BS_1} is the reflectivity of BS_1. The measurement positions for the DUT power fractions $P_{1,2}$ transmitting BS_1 are illustrated in Fig. 3.1.

In order to expose the sole heating effect of laser radiation incident on the DUT facet, optical injection from a heating laser is used. This laser is basically identical to the DUTs except for a slightly higher indium content to shift the emission spectrum by $20\,\text{nm}$ towards higher wavelengths. Furthermore the polarization of the heating laser is rotated by $90°$ by a half-wave plate ($\lambda/2$) to minimize the coupling to the DUT gain. The heating laser, as well, is collimated by L_1 and thus can be swept along the DUT facet by rotation of mirror M_2. Here, the same lens model L_1 is used for collimation and focusing of the heat laser so that the corresponding focus spot on the front facet has a lateral width of $100\,\mu\text{m}$ ($90\,\%$ p.c.). For the vertical direction, a nearly Gaussian distribution with a beam width of $1.8\,\mu\text{m}$ ($90\,\%$ p.c.) is calculated for the employed imaging system. Switching between feedback and injection can be done via the shutters S_1 and S_2.

The displacement $\Delta y = y - y_0$ of the return spots on the DUT facet is monitored by a vertical near-field imaging of the radiation by the heating laser or feedback that transmits the beam splitter BS_1. This near-field image is produced via lens L_5 with a magnification of 62.5 on a camera. The reference position of the QW ($y_0 = 0$) is identified by the maximum voltage drop of the laser voltage achievable by light injection [113].

The impact of back-irradiance is studied by destructive testing of the CO(M)D threshold current $I_{CO(M)D}$ and non-destructive thermoreflectance measurements. To prevent thermal rollover, $I_{CO(M)D}$ is obtained in short-pulse operation ($\tau = 4\,\mu s$, 0.02 % d.c.) by increasing I by 0.5 A till CO(M)D is triggered which manifests itself by a locally reduced near-field intensity accompanied by an irreversible power loss [100]. It is reported in [60] that the measurement of the CO(M)D threshold in short-pulse operation accelerates that type of degradation characteristic for long-term reliability tests in CW mode. Consequently, the short-pulse method can be used to assess the relative sensitivity to catastrophic damage of different devices also for CW operation.

In order to check the correlation between the occurrence of COMD and an enhanced facet temperature, a thermoreflectance setup is employed. In principle, the method of thermoreflectance is based on the change of facet reflectivity with temperature [114], which is monitored by a separate probe laser: For the present study, the probe laser is a fiber-coupled laser diode operating at 785 nm that is focused down with a power of 14 mW on the substrate with an $1/e^2$-diameter of 5 µm, at a position of $y = 15$ µm. The probe beam power reflected from the DUT facet P_{pb} and its change due to temperature variation ΔP_{pb} are then detected by a photodiode. Thereby, the change of facet temperature ΔT_f reads [115]

$$\Delta T_f = \underbrace{\left(\frac{1}{R_f} \frac{\partial R_f}{\partial T} \right)^{-1}}_{=:C_{TR}} \frac{\Delta P_{pb}}{P_{pb}}. \tag{3.2}$$

Since ΔP_{pb} is about eight orders of magnitude lower than the DUT radiation, the probe beam emission is filtered by a sequence of a grating monochromator and band-pass filters (similar to the setup in [116]) so that the spectrum emitted by the DUTs and the heating laser are suppressed by more than 100 dB. Furthermore, a lock-in detection is applied for the measurement of ΔP_{pb} and P_{pb} to improve the signal-to-noise ratio. Here, P_{pb} is measured with the probe laser operated in QCW mode (70 Hz, 50 % d.c.) and both the DUT and heating laser turned off. Afterwards ΔP_{pb} is recorded with the heating laser being incident on the turned-off DUT, operating in the same QCW mode as the probe laser before, which now is set to CW mode.

Moreover, the proportionality constant C_{TR} in equation (3.2) depends primarily on the semiconductor and coating material as well as the employed wavelength of the probe laser. The front facets of the employed DUTs are

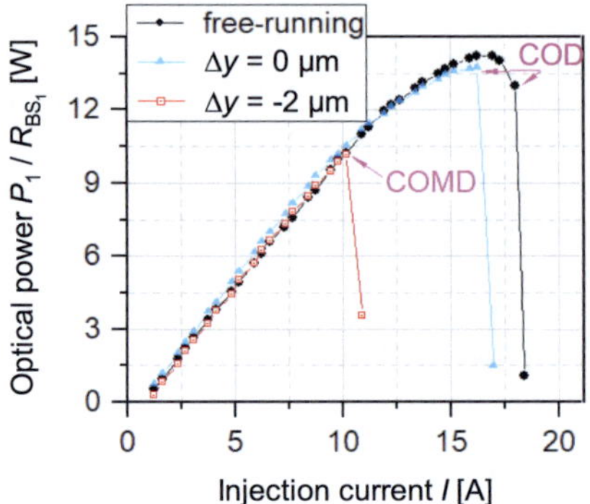

Fig. 3.2.: Device failure by misaligned feedback in CW operation: Exemplary L-I-curves of BALs ($L_i = 4\,\text{mm}$, $w_c = 100\,\mu\text{m}$, $R_b = 96\,\%$, $R_f = 2\,\%$) in external-cavity operation ($R_e = 4\,\%$) with different vertical displacements Δy of the feedback return spot on the front facet, compared to free-running operation. Negative values Δy denote a displacement towards the p-layers, [112].

coated by a two-layer coating that provides a reflectivity of $\sim 2\,\%$ for the wavelength range from 930 nm to 970 nm whereas at the probe wavelength, the reflectivity amounts to 15 %. Furthermore, for wavelengths around 785 nm the refractive index change of GaAs with temperature is about two orders of magnitude lower than for blue radiation around 450 nm [117], as it is frequently used for thermoreflectance setups [118]. This implies for the present study that C_{TR} is more sensitive to coating variations and therefore needs to be measured for each DUT, and, in particular, cannot be derived from the GaAs properties only, as done in [116]: Similar to a procedure reported in [119], C_{TR} is fit to a data set of P_{pb} values measured for different heat-sink temperatures, ranging from 20 to 50 °C, to a value of $C_{\text{TR}} = (-3.0 \pm 0.5) \times 10^3\,\text{K}^{-1}$. This numerical value is of the same order of magnitude as the values reported in [118] for GaAs.

3.1.2. CO(M)D threshold and front-facet temperature

At first, device failure is investigated for CW operation and a selected set of feedback configurations. The results are shown in Fig. 3.2. This graph manifests the threat of vertically misaligned feedback on the basis of exemplary single-device power characteristics: The L-I-curves for free-running and external-cavity operation ($R_e = 4\,\%$) without vertical displacement of the feedback spot ($\Delta y = 0$, matched feedback) show nearly the same power level before COD occurs. However, when the feedback spot is displaced by $2\,\mu\text{m}$ towards the p-side, the maximum achievable output power before the occurrence of COMD is 25 % less compared to the free-running case. Furthermore, the power curves for the free-running operation and the case with $\Delta y = -2\,\mu\mathrm{m}$ overlap for $I < I_{\mathrm{COMD}}$ without a significant difference visible. This observation suggests that the impact of the misaligned feedback is strongly localized and leaves the integrated device parameters determining the power slope unchanged. Note that the absolute power level achieved with the devices in free-running operation is rather small compared to data provided by other manufactures (e.g. [20]), which is due to the rather high thermal resistance of the cooling architecture employed in this study. The corresponding experimental setup is optimized for frequent change of DUTs so that the device on submount, as shown in Fig. 2.3, is clamped instead of soldered to the water-cooled heat sink, thereby trading a good thermal contact for mechanical flexibility.

To further investigate this assumption of localized heating, the CO(M)D threshold current $I_{\mathrm{CO(M)D}}$ and the front-facet temperature at a distance of $15\,\mu\mathrm{m}$ from the QW in the substrate ΔT_{f} is measured with the heating laser spot swept along the vertical position of the DUT facet. This temperature measurement is conducted for both p-side down and p-side up mounting of the chip on the submount. Therefore, the obtained facet-temperature rise is defined as $\Delta T_{\mathrm{f}} := T(x = 0, y = \pm 15\,\mu\mathrm{m}, z = 4\,\mathrm{mm}) - T_{\mathrm{hs}}$, where "+" refers to p-down and "−" for p-up mounting. Furthermore, the power of the heating laser spot on the DUT facet amounts to $1\,\mathrm{W}$ and $0.5\,\mathrm{W}$ for p-down and p-up mounting, respectively. Regarding an operation point with $10\,\mathrm{W}$ optical output power, these heat powers are equivalent to a feedback strength of 10 % and 5 %, respectively, which are common for single-side emitting EC-BALs [31,96]. In the latter case, the reduction of heating power is employed to prevent damage of the DUT, for the thermal resistance of the p-up mounted devices is more than two times higher compared to DUTs mounted p-down. The corresponding results are shown in Fig. 3.3 in comparison with the data obtained for free-running, short-pulse operation. Here, the curve for $I_{\mathrm{CO(M)D}}$ represents the average result of a multi-device study of at least four DUTs for each configuration, while the data for ΔT_{f} originate from representative single devices.

At first, the mean $I_{\mathrm{CO(M)D}}$ in free-running operation is found to be $\sim 57\,\mathrm{A}$. As a consequence, for the DUTs to be investigated with external injection from the heating laser, it is first assured that they can tolerate a current of $50\,\mathrm{A}$ in free-running mode to separate out pre-damaged material. Figure 3.3 shows that the lowest values for $I_{\mathrm{CO(M)D}}$, which are less than 25 % of the free-running value, can be found for the heating laser displaced 2 to $3\,\mu\mathrm{m}$ towards the p-side. Furthermore, it is exactly these positions where an actual front-facet damage is visible. Consequently, for all other configurations (including free-running mode), the damage origin must lie inside the bulk material, which is investigated in detail at a later point. As a result, the interface region be-

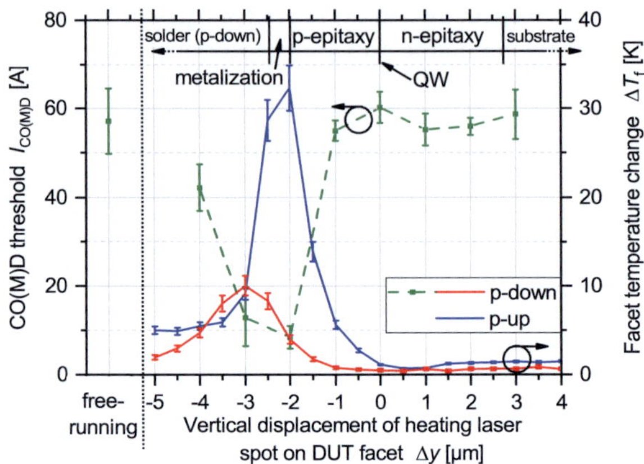

Fig. 3.3.: Device failure by misaligned feedback: The CO(M)D current threshold is measured in short-pulse mode ($\tau = 4\,\mu s$, $0.02\,\%$ d.c.) for single-side emitting devices ($L_i = 4\,mm$, $w_c = 100\,\mu m$, $R_b = 96\,\%$, $R_f = 2\,\%$, p-down mounting) with the heating laser irradiating the DUT facet with a power of $1\,W$. For the corresponding thermoreflectance measurement, the front-facet temperature ΔT_f is measured on the substrate at a distance of $15\,\mu m$ to the QW while the DUT is turned off. The heating laser power incident on the DUT facet for the measurement of ΔT_f amounts to $1\,W$ and $0.5\,W$ for p-down and p-up mounted devices, respectively. Published in excerpts in [112].

tween semiconductor p-side and the adjacent metal regions are the weak spots of this kind of laser diode with respect to misaligned back-irradiance. On the contrary, irradiation of the waveguide, the highly n-doped layers as well as the substrate with a power of $1\,W$ at $970\,nm$ does not lead to a significant change of $I_{CO(M)D}$ compared to the value obtained in free-running operation.

In order to understand these results of destructive testing and potentially correlate them with front-facet heating the curves of $I_{CO(M)D}$ and ΔT_f for devices mounted p-side down shown in Fig. 3.3 are compared in the following.

This comparison reveals that an enhanced ΔT_f causes a decrease in $I_{CO(M)D}$. However, the peak positions of $I_{CO(M)D}$ and ΔT_f are displaced relative to one another by $\sim 1\,\mu m$. This leads to the asymmetry that for the displacements $\Delta y = -2, -4\,\mu m$ almost the same facet temperature is measured, for the damage thresholds, however, one finds $\Delta I_{CO(M)D}(-2\,\mu m) < 0.25 \cdot I_{CO(M)D}(-4\,\mu m)$.

To understand the observed deviation between thermoreflectance and damage threshold data, the different metal surfaces adjacent to the semiconductor are analyzed in more detail by means of the comparison of ΔT_f for p-down and p-up mounting: Since for p-up devices the solder is attached to the n-side, with these devices the p-side metalization can be irradiated separately without heating up parts of the solder.

In Fig. 3.3 it can be seen that, for p-up mounting, the facet temperature is highest when large parts of the heating laser beam cover the p-side metalization, i.e. for $\Delta y = -2$ to $-2.5\,\mu m$ where the heating laser spot is centered at the p-side metalization. For $\Delta y < -2.5\,\mu m$, ΔT_f drops rapidly for p-up-mounting since there is no solder nearby. For p-down mounting, however, the maximum of ΔT_f is located at $\Delta y = -3\,\mu m$, i.e. the solder layer, where the spot of the heating laser hits metal material exclusively. Furthermore, the fact that the peak values of ΔT_f are higher for p-up than for p-down mounting is mainly attributed to the difference in the thermal resistances of these two configurations.

These observations lead to the following two important conclusions:

- The front-facet temperature is increased the most by irradiation of metal-based layers where the radiation is basically absorbed at the surface due to the small penetration depth of typically $<1\,\mu m$ for Au. The effective absorption of laser light at the AuSn solder alloy is enhanced by its porous surface [120] that favors multiple reflections of an incoming beam leading to diffuse back-scattering: This argument is confirmed by a simple reflectivity measurement with a witness sample of the employed AuSn solder that revealed a reflectivity for targeted reflexion of only $\sim25\,\%$ at wavelength of $940\,nm$ for close-to normal incidence.

  After the first publication of the impact front-facet heating via solder irradiation by the author of this thesis in 2018 [112], this effect has been subsequently confirmed for BALs operating in the 800, 900 and 1000 nm regime [121].

- The CO(M)D threshold is decreased by elevated facet temperatures. However, no monotonic relation between these two parameters could be found. Instead, an irradiation of both highly p-doped cap layers and the adjacent metalization leads to the lowest $I_\mathrm{CO(M)D}$ but only a medium rise of ΔT_f. These highly p-doped layers at the end of the epitaxial layer stack exhibit an free-carrier absorption coefficient in the order of $1000\,cm^{-1}$ which is equivalent to a penetration depth of $\sim10\,\mu m$.

  This indicates that absorption close to the longitudinal interface of the electrically pumped and non-pumped sections (cf. Fig. 2.4b) supports the occurrence of COMD.

In order to check this thesis, in the following the longitudinal origin of the catastrophic damage is investigated by cathodoluminescence for the devices from Fig. 3.3.

3.1.3. Longitudinal origin of COD

Regions in a semiconductor laser that suffered from COD exhibit a damaged crystal structure and a spatially changed mixing ratio of the employed semiconductor alloys [122] and hence changed luminescence properties. That means in particular that the characteristic emission lines of the intact material are strongly suppressed in case of COD. The method of cathodoluminescence microscopy utilizes the changed emission behavior to identify damaged material: The epitaxial layers are excited by a high-energy ($\sim$30 keV) electron beam and the subsequent luminescence is spatially and spectrally resolved.

For the DUTs tested subject to external laser heating (cf. Fig. 3.3), the damage pattern can be subdivided into those devices that exhibit bulk damage (COD) and those showing visible facet failure (COMD) which could be identified for $\Delta y = 2 .. 3\,\mu$m only. For these two patterns representative panchromatic cathodoluminescence images of the front section are depicted in Fig. 3.4.

There, the damaged sections manifest themselves via black DLDs. In the exemplary cathodoluminescence image in Fig. 3.4a, the DLDs start at 43 μm behind the front facet. For the remaining devices that suffered from COD, the distance of the origin of bulk damage to the front facet ranges from 30 to 100 μm. In case of COMD, as exemplarily depicted in Fig. 3.4b, the DLDs begin evidently right at the facet and spread towards the bulk. With respect to the lateral direction, the origin of the damage nearly coincides with the position of the highest intensity filaments visible in the near field (cf. Fig. 3.4, top). Furthermore, the longitudinal direction of defect propagation (towards the laser back side) is revealed by the increasing number of branches of the DLDs during spreading [123]. Thus the defect lines fill nearly the full lateral contact width and extend longitudinally to about half the cavity (not shown in Fig. 3.4). Furthermore, from spectrally resolved versions of the cathodoluminescence images shown in Fig. 3.4 it can be drawn that the damaged volume is confined to the waveguide and QW. Especially, there are no DLDs in the cladding layers.

The observations that the longitudinal origin position of CO(M)D depends on the vertical displacement of back-irradiance and that bulk damage starts close to the interface of pumped and non-pumped area suggests a correlation to the corresponding longitudinal temperature distributions. Therefore, the steady-state vertical and longitudinal temperature profiles are analyzed in the following for an exemplary operation point of a DUT with and without external laser heating on the solder by means of the SEMSIS thermal solver.

Theoretical studies modeling the path to CO(M)D emphasize the necessity of temperature-dependent heat sources [61] and thermal conductivities [57] and current redistribution [58] that enable the thermal runaway at a local hotspot where a certain critical temperature is reached. As a consequence of

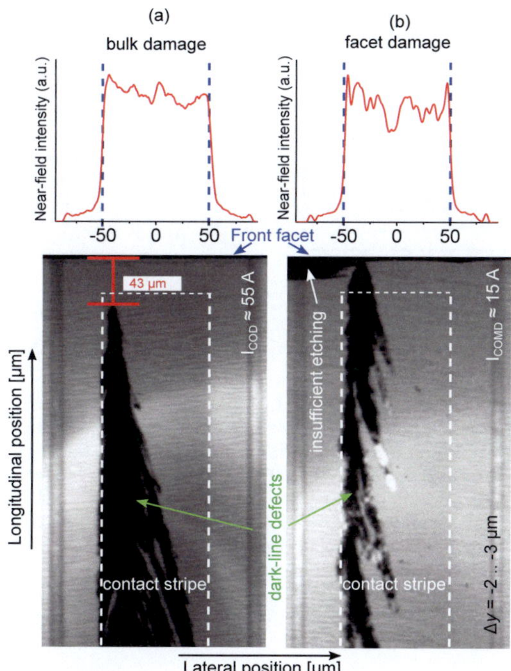

Fig. 3.4.: Front-facet near-field (top) obtained a current step of 0.5 A prior to device failure and corresponding lateral-longitudinal panchromatic cathodoluminescence images after catastrophic damage (bottom): Representative measurements for DUTs from the destructive testing illustrated in Fig. 3.3 are shown for bulk (a) and facet (b) damage. Damaged areas are characterized by dark-line defects, [112].

that understanding, a steady-state analysis provided by the SEMSIS package where constant injection current and thermal conductivities are chosen cannot reproduce thermal runaway and temperatures as high as the melting point of waveguide or QW. A second consequence from these theoretical reports is, however, that the steady-state solution of the heat equation (2.17), $T(x, y, z)$, as provided by SEMSIS can reveal local temperature hotspots that are potential seeds for a thermal runaway.

Therefore a BAL as used for the thermoreflectance measurements and destructive testing depicted in Fig. 3.3 is simulated by the SEMSIS package at an exemplary operation point providing the following performance parameters obtained from experimental measurements (cf. Fig. 3.2): $I = 14$ A, $P_{\mathrm{opt}} = 13.25$ W, $\eta_{\mathrm{eo}} = 59\%$. Here, especially the non-pumped sections with $L_{\mathrm{un}} = 30$ µm are considered.

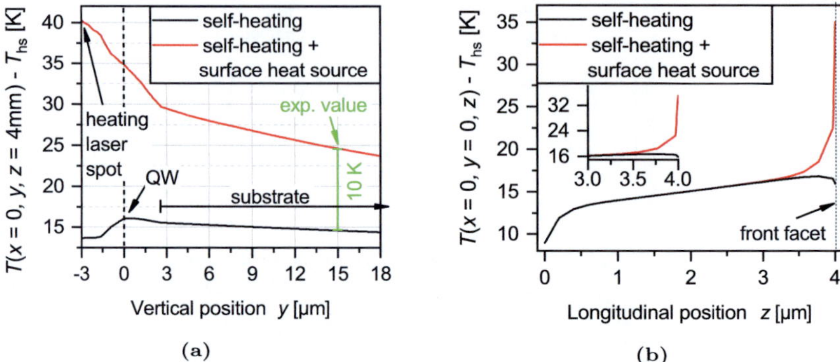

(a)　　　(b)

Fig. 3.5.: Temperature inside a BAL ($L_i = 4\,\text{mm}$, $w_c = 100\,\mu\text{m}$, $R_b = 96\,\%$, $R_f = 2\,\%$, p-down mounting) for sole laser self-heating and an additional surface heat source at $y = -3\,\mu\text{m}$ which models solder heating with an external laser source, simulated with the SEMSIS package: Vertical profile at the front facet and the center of the contact stripe (a) and longitudinal profile at the QW position centered at the contact stripe (b). The strength of the surface heat source is chosen in a way that the simulation reproduces the experimental value measured at $y = 15\,\mu\text{m}$ on the substrate, shown in Fig. 3.3. Published in [112].

To include the effect of external laser heating, a surface heat source with the lateral-vertical dimensions of the beam of the heating laser at the DUT facet (cf. subsection 3.1.1) is centered at $\Delta y = -3\,\mu\text{m}$: In vertical direction a Gaussian distribution with a diameter of $1.8\,\mu\text{m}$ at $90\,\%$ p.c. is chosen and in the lateral direction the surface heat source is distributed homogeneously over the contact opening width. Since this simulation analysis is focused on a relative temperature comparison between free-running and operation with external laser heating, the bottom of the submount is kept at T_{hs}, which leads to an underestimation of the absolute temperature. Additionally, the amplitude of the surface heat source is chosen in a way that the front-facet temperature rise due to external heating amounts to $10\,\text{K}$ at $y = 15\,\mu\text{m}$ as experimentally obtained for $\Delta y = -3\,\mu\text{m}$ in Fig. 3.3. The corresponding results are shown in Fig. 3.5.

A vertical temperature profile at the front facet at the center of the contact stripe $T(x = 0, y, z = 4\,\text{mm}) - T_{\text{hs}}$ is depicted in Fig. 3.5a. Here, it is evident that by activating external laser heating on the solder the peak temperature position is shifted from the QW to the center of the heating beam. Additionally, the temperature increase at the QW amounts to almost $20\,\text{K}$ which is about twice the value measured $15\,\mu\text{m}$ above the QW on the substrate with the DUT turned off.

Furthermore, in Fig. 3.5b the corresponding longitudinal profile $T(x = 0, y = 0, z) - T_{hs}$ illustrates the impact of heating the solder surface along the cavity axis: For sole laser self-heating, the temperature slightly decreases on the last $\sim$100 µm towards the front facet due to the presence of the non-pumped section. By implication, the temperature peak is present inside the bulk about where the origin of the DLDs was found in the cathodo-luminescence analysis depicted in Fig. 3.4a. With an additional surface heat source on the solder, the temperature sharply increases on the last $\sim$500 µm towards front which explains why for the corresponding devices also the origin of catastrophic damage is located at the facet.

Hence the thermal simulation in Fig. 3.5 explains the experimental origins of CO(M)D shown in Fig. 3.4 by analysis of the longitudinal position of the maximum steady-state temperature.

In summary, for the InGaAs/AlGaAs-based structure investigated, irradiation of the outer, highly p-doped layers in combination with the adjacent metal layers leads to strong temperature increase close to the front facet which reduces the threshold for catastrophic device failure significantly and shifts the damage origin from inside the bulk to the emission facet. Furthermore, especially for matched optical feedback of 4 % no essential reduction of the CW damage threshold is observed.

Therefore it is important for the construction of EC-BALs to mitigate displacement or expansion of the feedback spot towards the p-side by the optical and thermo-mechanical design of the external-cavity configuration.

3.2. Lateral displacement

In this section, the impact of lateral displacement of back-irradiance is studied with regard to changes of beam quality. Such a study is particularly important for rather complex external-cavity configurations used for DWBC [12] to enable root-cause identification of beam quality deterioration.

In the previous section 3.1, the investigation of vertical beam displacement focused on the absorption of the misaligned radiation while the accompanying change in coupling efficiency quantified by Γ_c was neglected. This is because non-ideal back-coupling in the vertical direction does not change the time-averaged vertical near field due to strong index-guiding [70]. In the lateral direction, however, the emission characteristics of the employed gain-guided BAL structure are determined by the lateral-longitudinal profiles of the gain and the thermally and carrier-induced index changes, which are dependent on the spatial and angular distribution of the back-coupled feedback intensity.

For this investigation, an asymmetric EC-BAL with $R_b = 96\,\%$, $R_f = 0.5\,\%$ and $R_e = 5.2\,\%$ is compared to a symmetric EC-BAL with $R_b = 4.5\,\%$,

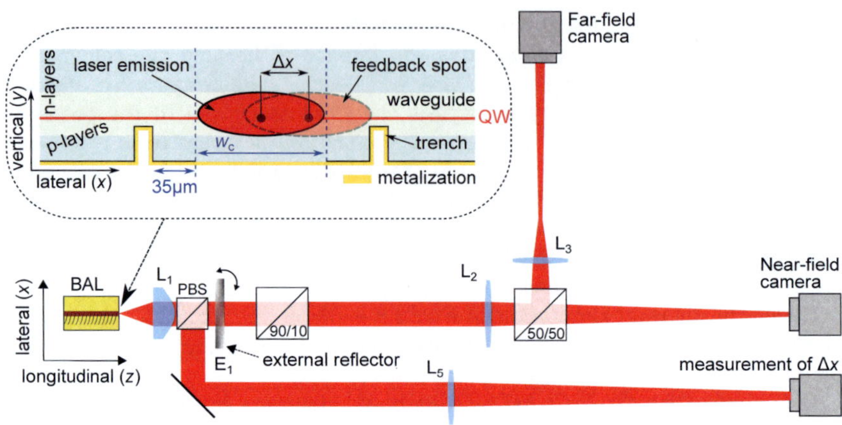

Fig. 3.6.: Experimental setup used for beam quality characterization of BALs subjected to laterally displaced optical feedback. The residual TE reflectivity of a PBS is used to detect the lateral displacement Δx of the centers of the laser emission and the feedback spot on the DUT facet.

$R_\text{f} = 0.5\,\%$ and $R_\text{e} = 5.2\,\%$. Both designs exhibit a contact opening of $130\,\mu\text{m}$, an internal cavity length of $L_\text{i} = 5\,\text{mm}$, and a 2f external cavity with a length of $L_\text{e} \approx 20\,\text{mm}$.

Figure 3.6 depicts the corresponding experimental setup which is basically the same as shown in Fig. 2.2 extended by an option to measure the lateral displacement Δx of the feedback beam: A PBS is placed between the collimation lens L_1 and the external reflector E_1 which separates transverse-magnetic (TM) from TE emission. For the reflected beam, the extinction ratio of TM over TE reflectivity is only about 60:1 so that about $1.6\,\%$ of the TE power back-reflected by E_1 is imaged via lens L_5 with a magnification of 62.5 onto a camera and is used to measure the lateral displacement Δx between the centers of the emission and the feedback spot on the DUT facet. The definition of Δx is illustrated in the inset of Fig. 3.6. The reference position $\Delta x = 0$ manifests itself experimentally by the maximum achievable voltage drop across the laser diode for a fixed feedback strength. In order not to cover the metalization inside the etched trench gap, which would cause strong facet heating according to results of the previous section 3.1, the displacement is restricted to $|\Delta x| \leq 30\,\mu\text{m}$ in this study, which, for the employed external cavity, corresponds to a maximum rotation of E_1 by $\pm 2\,\text{mrad}$. A corresponding pointing of the back-coupled light at the DUT facet is below $1\,\text{mrad}$ and thus negligible compared to the size of θ_{95}.

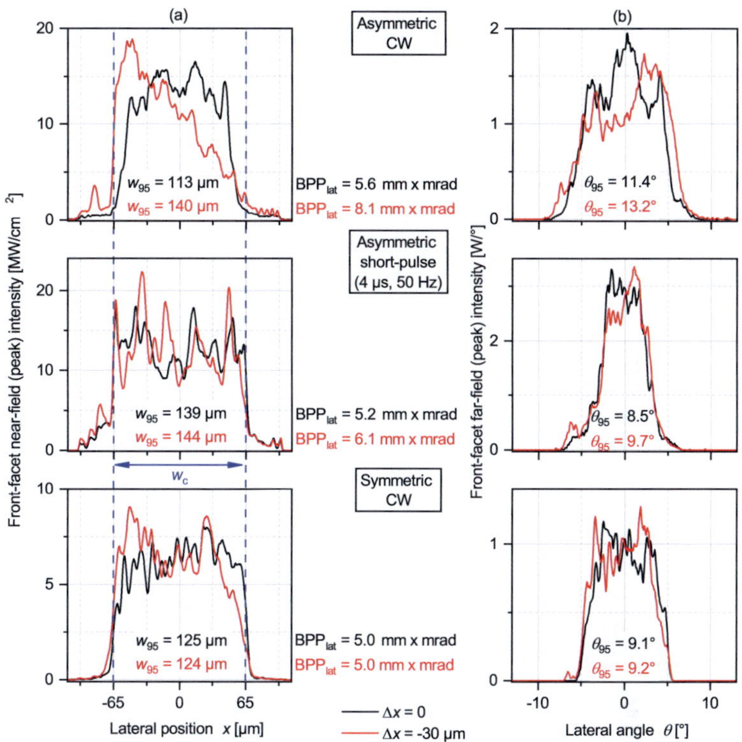

Fig. 3.7.: Experimental near-(a) and far-field(b) profiles for asymmetric ($R_b = 96\,\%$, $R_f = 0.5\,\%$, $R_e = 5.2\,\%$) and symmetric ($R_b = 4.5\,\%$, $R_f = 0.5\,\%$, $R_e = 5.2\,\%$) EC-BAL configurations for $\Delta x = 0$, $-30\,\mu$m and $I = 18\,$A.

In the Figs. 3.7a and b the lateral near- and far-field profiles, respectively, are shown for the asymmetric EC-BAL in short-pulse ($\tau = 4\,\mu$s, $0.02\,\%$ d.c.) and CW operation and for the symmetric EC-BAL in CW mode. In order to include the effect of near-field narrowing and accompanying enhanced lateral SHB, the corresponding drive current is set to $18\,$A for all measurements in this figure, since at this current level near-field narrowing is present even for $140\,\mu$m-wide BALs as shown in Fig. 2.13.

From Fig. 3.7 it is evident that the impact of a lateral feedback misalignment of $\Delta x = -30\,\mu$m is most dominant for the asymmetric EC-BAL in CW operation. Here, the feedback displacement causes the near-field width w_{95} to increase by almost $30\,\mu$m and the corresponding profile to deform: Compared to the case of no displacement, for $\Delta x = -30\,\mu$m the near-field intensity increases

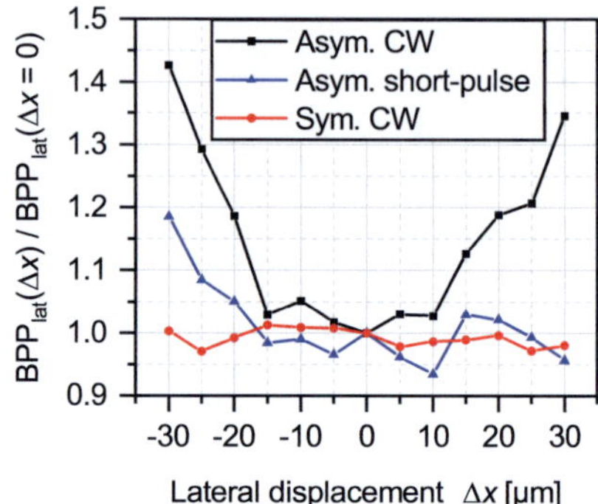

Fig. 3.8.: Experimentally measured, relative change of BPP$_{lat}$ as a function lateral displacement of the feedback spot on the DUT front facet: Experimental comparison of an asymmetric ($R_b = 96\,\%$, $R_f = 0.5\,\%$, $R_e = 5.2\,\%$) with an symmetric ($R_b = 4.5\,\%$, $R_f = 0.5\,\%$, $R_e = 5.2\,\%$) EC-BAL configuration with $w_c = 130\,\mu\text{m}$ and $L_i = 5\,\text{mm}$ at $I = 18\,\text{A}$. Short-pulse parameters are $\tau = 4\,\mu\text{s}$ and $0.02\,\%$ d.c.

in the vicinity of the contact edge to which the feedback spot is displaced and decreases at the opposite contact edge. For this amount of displacement, the corresponding far-field width θ_{95} additionally increases by almost $2°$ which results in a deterioration of BPP$_{lat}$ by $45\,\%$ from 5.6 to $8.1\,\text{mm} \times \text{mrad}$.

Compared with the asymmetric EC-BAL in CW operation, the impact of feedback displacement of $-30\,\mu\text{m}$ is much more reduced in short-pulse mode and for the symmetric EC-BAL in CW mode: For the pulsed EC-BAL, where thermal lensing is negligible ($T_{jun} - T_{hs} \sim 5\,\text{K}$, cf. Fig. 2.17), feedback displacement leaves the near-field centroid position basically unchanged, but causes a rearrangement of filaments. As a result, the corresponding BPP$_{lat}$ increases by only $17\,\%$ which is due to changes of side lobes of the near- and far-field profile. Especially for the symmetric configuration, which exhibits reduced side lobes beside the contact edges due to reduced lateral carrier accumulation (cf. Fig. 2.15), the beam parameters and hence BPP$_{lat}$ remain basically unaltered.

These observations indicate that beam quality degradation due to feedback misalignment is supported by lateral carrier accumulation as well as strong thermal lensing leading to near-field narrowing and accompanying lateral SHB. To confirm that this conclusion is generally valid, in Fig. 3.8 the relative change of BPP$_{lat}$ is depicted as a function of Δx with reference to $\Delta x = 0$. This graph shows for CW operation that while for the symmetric configuration the BPP$_{lat}$ varies only in range of less than $\pm 5\,\%$, the asymmetric laser exhibits a strong deterioration of BPP$_{lat}$ beyond $30\,\%$ towards $\Delta x = \pm 30\,\mu\text{m}$. This fundamental difference with regard to beam quality degradation due to feedback misalignment manifests a further advantage of symmetric as opposed to asymmetric EC-BALs.

If however the lateral displacement can be restricted to $\Delta x = \pm 10\,\mu\text{m}$, the comparison between short-pulse and CW operation in Fig. 3.8 demonstrates that the deteriorating impact of strong thermal lensing is small in the sense that BPP_{lat} varies in the same range ($\pm 7\,\%$) as in short-pulse mode.

Furthermore, 4f external-cavity configurations can help reduce the sensitivity for spatial feedback displacement, since a tilt of the respective external reflectors does not lead to a spatial misalignment of the feedback spot on the DUT front facet.

4. Conclusion

In this thesis the impact of the internal longitudinal temperature and intensity distributions on the lateral emission characteristics of gain-guided, InGaAs-based BALs subject to external optical feedback has been investigated. For matched optical feedback the main findings apply likewise for free-running and external-cavity operation. They are:

- By means of numerical simulations, it has been demonstrated that the narrowing of the near field at the front facet in the range of 20 % in single-side emitting BALs at high injection currents, which is revealed to be accompanied by a far-field blooming, an increased lateral spatial hole burning, and back-facet near-field blooming, can be attributed to strong lateral thermal lensing that increases from the back to the laser front.

- By implication, a homogenization of the longitudinal device temperature can be used to avoid the shrinkage of the front-facet near field and thus potentially enhance the laser lifetime while the lateral beam parameter product deviates only by $<10\,\%$.

- In an experimental multi-device study, it could be shown that the homogenization of the longitudinal intensity and thus temperature profile by means of symmetrized facet reflectivities can approximate the case of longitudinal isothermalty sufficiently in order to mitigate front-facet near-field narrowing at high powers: Throughout the investigated current range, the experimental near-field widths including 95 % p.c. showed a maximum deviation of only 10 % from the width of the contact opening.

- Additionally, for the whole investigated current range, these double-side emitting BALs with a contact-stripe width of $140\,\mu m$ exhibit a decreased BPP_{lat} by about $1\,mm \times mrad$ compared to unidirectional devices with similar mirror loss. This improvement in terms of beam quality has been shown to be due to an decreased lateral carrier accumulation because of reduced internal intensity for the bidirectional devices.

- Based on these results, double-side emitting wavelength-stabilized external-cavity BALs, that outperform the equivalent single-side emitting configurations simultaneously in terms of total optical power, electro-optical efficiency, lateral beam quality as wells as reliability and locking range, are feasible.

For mismatched optical feedback, lateral and vertical displacements of the feedback spots have been found to influence beam quality and lifetime through the following main mechanisms:

- Focused irradation of the solder and metalization along with the adjacent highly p-doped semiconductor cap layers causes a strong facet heating. This causes the origin of COD, which for no additional facet heating is located $\sim100\,\mu$m behind the facet due to the use of non-pumped mirrors, to appear right at the facet as COMD. Here, solder heating by a laser beam with a power of 1 W is observed to lower the threshold for CO(M)D by 75 % compared to the free-running value.

- Furthermore, it has been demonstrated that displacing the feedback return spot laterally by $\pm30\,\mu$m can cause a degradation of lateral beam quality of gain-guided single-side emitting DUTs with a contact width of $100\,\mu$m by over 40 %. Such a beam quality degradation could be reduced by symmetric outcoupling to <5 %.

Since the double-side emitting device architecture requires an elaborate mechanical design as well as additional optics to collimate and combine the two beams, and thus higher expenses, laser diode manufactures might rather want to stay with unidirectional devices. For this situation a future study may investigate further means for longitudinal temperature homogenization: For example a reduced internal absorption by a tailored epitaxial design can decrease longitudinal temperature gradients. Additionally, a tapered contact, similar to the approach shown in [69], can help to spread the waste heat more homogeneously: The heat source density could be kept approximately uniform if the lateral stripe width is increased towards the front facet to the same extent as the longitudinal light intensity profile. Furthermore, tailoring the depth and the lateral distance between built-in index trenches may help to reduce the sensitivity of BALs to beam quality deterioration by both thermal lensing and lateral feedback misalignment.

Appendices

A. Simulation parameters

Two-dimensional simulation: BALaser and SEMSIS

In this section, the numerical parameters used for the simulations with the extended BALaser package and SEMSIS are listed in Tab. A.1.

One-dimensional simulation

Furthermore, for one-dimensional calculations of $S(z)$ and $N(z)$ based on the model in [88], the numerical parameters are summarized in Tab. A.2. Basically, this one-dimensional model can be deduced from the two-dimensional equations (2.11) and (2.12) disregarding all t- and x-dependent terms as well as real refractive index changes. The parameters are similar to those used with the BALaser package neglecting the dependences on temperature.

Lumped model

Moreover, for modeling the e-o efficiency η_{eo} a simple lumped model similar to the one presented in [17] is used. Here η_{eo} is modeled as a function of α_{m} as follows

$$\eta_{\mathrm{eo}}(\alpha_{\mathrm{m}}) = \frac{S_0 \frac{\alpha_{\mathrm{m}}}{\alpha_{\mathrm{i}}+\alpha_{\mathrm{m}}} \left[I - I_{\mathrm{th}}(N, \Delta T, \alpha_{\mathrm{m}})\right]}{I\left(U_{\mathrm{F}}(N(\Delta T, \alpha_{\mathrm{m}})) + 0.02\,\mathrm{V} + R_{\mathrm{s}}I\right)}. \tag{A.1}$$

Furthermore, the threshold current I_{th} and scalar carrier density in equation (A.1) are derived from the rate of non-radiative recombination and the a logarithmic gain model:

$$I_{\mathrm{th}}(N(\Delta T, \alpha_{\mathrm{m}})) = L_{\mathrm{i}} w_{\mathrm{c}} e d_{\mathrm{qw}} \left(AN + BN^2 + CN^3\right) \tag{A.2}$$

$$N(\Delta T, \alpha_{\mathrm{m}}) = N_{\mathrm{tr}}(\Delta T) \exp\left(\frac{\alpha_{\mathrm{i}} + \alpha_{\mathrm{m}}}{g_0(\Delta T)}\right). \tag{A.3}$$

The parameters used for calculations based on this lumped model are listed in Tab. A.3. The values for S_0, g_0 and ΔT are extracted from (manual) fits to the power curve and the junction temperature as depicted in the Figs. 2.13b and c. The model for U_F is the same as in Tab. A.1.

Tab. A.1.: Numerical values for the parameters used with the BALaser and SEMSIS simulation software.

	BALaser	SEMSIS [87]	unit
$\bar{n}$	3.42	3.42	
n_g	3.9	–	
A	2.2×10^8	2×10^8	s^{-1}
B	$1.7 \times 10^{-16}(300/\overline{T}_\mathrm{QW})$ [124]	1.4×10^{-16} [40]	m^3/s
C	$4.2 \times 10^{-42}\sqrt{\frac{\overline{T}_\mathrm{QW}}{468.4\,\mathrm{K}}}$ $\times \exp\left(\frac{468.4\,\mathrm{K}}{300\,\mathrm{K}} - \frac{468.4\,\mathrm{K}}{\overline{T}_\mathrm{QW}}\right)$ [124]	3.5×10^{-42}	m^6/s
D_N	26×10^{-4}	19.6×10^{-4}	m^2/s
g_0	$1685.0 - 4.1\,\mathrm{K}^{-1}(\overline{T}_\mathrm{QW} - 300\,\mathrm{K})$	$1586.20 - 0.45\,\mathrm{K}^{-1}$ $\times (T(x,0,z) - T_\mathrm{hs})$	m^{-1}
N_tr	$(1.439 + 0.007\,\mathrm{K}^{-1}$ $\times(\overline{T}_\mathrm{QW} - 300\,\mathrm{K})) \times 10^{24}$	$(2.00 + 0.01\,\mathrm{K}^{-1}$ $\times(T(x,0,z) - T_\mathrm{hs})) \times 10^{24}$	m^{-3}
α_i	$\begin{cases} 90 & \text{for } I \leq 11.8\,\mathrm{A} \\ 100 & \text{else} \end{cases}$	–	m^{-1}
α_0	–	20	m^{-1}
σ_n	4×10^{-22}	3×10^{-22}	m^2
σ_p	12×10^{-22}	7×10^{-22}	m^2
Δn_0	-9.7×10^{-3}	-3.84×10^{-3}	
$\frac{\partial n}{\partial T}$	3×10^{-4} [21]	4×10^{-4}	K^{-1}
n'	$(5.120 + 0.019\,\mathrm{K}^{-1}$ $\times(\overline{T}_\mathrm{QW} - 300\,\mathrm{K})) \times 10^{-32}$	1.7×10^{-26}	m^3
R_s	15.5	–	$\mathrm{m}\Omega$
b_F	$-1.563\,71 \times 10^{-6}$ $+2.222\,91 \times 10^{-11}$ $\times \exp(0.04344 \cdot \overline{T}_\mathrm{QW})$	–	V
c_F	$-0.023\,71$ $+0.652\,99$ $\times \exp(-0.00367 \cdot \overline{T}_\mathrm{QW})$	–	

Tab. A.2.: Numerical values for the parameters used with the one-dimensional model in [88].

parameter	1D model	unit
g_0	1700	m^{-1}
N_{tr}	1.44×10^{24}	m^{-3}
A	3×10^8	s^{-1}
B	1.7×10^{-16}	m^3/s
C	3.4×10^{-42}	m^6/s
α_{i}	60	m^{-1}

Tab. A.3.: Numerical values for the parameters used with lumped model of the equations (A.1)–(A.3).

parameter	lumped model	unit
g_0	$1400.0 - 0.4\,\mathrm{K}^{-1}\Delta T$	m^{-1}
N_{tr}	$(1.44 + 0.01\,\mathrm{K}^{-1}\Delta T) \times 10^{24}$	m^{-3}
A	2.2×10^8	s^{-1}
B	1.7×10^{-16}	m^3/s
C	3.4×10^{-42}	m^6/s
α_{i}	60	m^{-1}
S_0	1.32	$\mathrm{W\,A}^{-1}$
R_{s}	15	$\mathrm{m}\Omega$
w_{c}	140	$\mu\mathrm{m}$
L_{i}	5	mm
ΔT	$5.3 + 0.3547\,\mathrm{A}^{-1}I + 0.0745\,\mathrm{A}^{-2}I^2$	K
a_{F}	$1.72009 - 0.2753\exp\left(0.00373(\Delta T + 300)\right)$	V

Bibliography

[1] P. Crump, K. Hasler, H. Wenzel, S. Knigge, F. Bugge, and G. Erbert, "High efficiency, 8W narrow-stripe broad-area lasers with in-plane beam-parameter-product below 2 mm mrad," in *2013 Conference on Lasers Electro-Optics Europe International Quantum Electronics Conference CLEO EUROPE/IQEC*, pp. 1–1, May 2013.

[2] P. Crump, J. Wang, T. Crum, S. Das, M. DeVito, W. Dong, J. Farmer, Y. Feng, M. Grimshaw, D. Wise, and S. Zhang, "> 360W and > 70% efficient GaAs-based diode lasers," in *Proc. SPIE*, vol. 5711, pp. 5711 – 5711 – 9, 2005.

[3] P. Crump, M. Grimshaw, J. Wang, W. Dong, S. Zhang, S. Das, J. Farmer, M. DeVito, L. S. Meng, and J. K. Brasseur, "85% power conversion efficiency 975-nm broad area diode lasers at -50°C, 76% at 10°C," in *2006 Conference on Lasers and Electro-Optics and 2006 Quantum Electronics and Laser Science Conference*, pp. 1–2, May 2006.

[4] A. Knigge, G. Erbert, J. Jonsson, W. Pittroff, R. Staske, B. Sumpf, M. Weyers, and G. Tränkle, "Passively cooled 940 nm laser bars with 73% wall-plug efficiency at 70 W and 25°C," *Electron. Lett.*, vol. 41, pp. 250–251, March 2005.

[5] P. Crump, G. Erbert, H. Wenzel, C. Frevert, C. M. Schultz, K. Hasler, R. Staske, B. Sumpf, A. Maaßdorf, F. Bugge, S. Knigge, and G. Tränkle, "Efficient high-power laser diodes," *IEEE J. Sel. Topics Quantum Electron.*, vol. 19, pp. 1501211–1501211, July 2013.

[6] P. Leisher and R. Martinsen, "A penalty-free approach to wavelength stabilization of high power diode lasers from 900 nm to 1900 nm," in *22nd IEEE International Semiconductor Laser Conference*, pp. 146–147, Sept 2010.

[7] R. L. Byer, "Diode laser-pumped solid-state lasers," *Science*, vol. 239, no. 4841, pp. 742–747, 1988.

[8] K. Kincade, A. Nogee, G. Overton, D. Belforte, and C. Holton, "Annual laser market review & forecast: Lasers enabling lasers," *Laser Focus World*, Jan. 2018.

[9] D. Blázquez-Sánchez, B. Weichelt, A. Austerschulte, A. Voss, T. Graf, A. Killi, H.-C. Eckstein, M. Stumpf, A. L. Matthes, and U. D. Zeitner, "Improving the brightness of a multi-kilowatt single thin-disk laser by an aspherical phase front correction," *Opt. Lett.*, vol. 36, pp. 799–801, Mar 2011.

[10] C. Jauregui, J. Limpert, and A. Tünnermann, "High-power fibre lasers," *Nature Photonics*, vol. 7, pp. 861–867, 10 2013.

[11] R. K. Huang, B. Chann, and J. D. Glenn, "Ultra-high brightness wavelength-stabilized kW-class fiber coupled diode laser," in *Proc. SPIE*, vol. 7918, pp. 7918 – 7918 – 9, 2011.

[12] M. Haas, S. Rauch, S. Nagel, R. Beißwanger, T. Dekorsy, and H. Zimer, "Beam quality deterioration in dense wavelength beam-combined broad-area diode lasers," *IEEE J. Quantum Electron.*, vol. 53, pp. 1–11, June 2017.

[13] S. Ried, S. Rauch, L. Irmler, J. Rikels, A. Killi, E. Papastathopoulos, E. Sarailou, and H. Zimer, "Next generation diode lasers with enhanced brightness," in *Proc. SPIE*, vol. 10514, pp. 10514 – 10514 – 8, 2018.

[14] "Communications: Laser Technik Journal 2/2016," *Laser Technik Journal*, vol. 13, no. 2, pp. 14–16, 2016.

[15] M. Hempel, M. Chi, P. M. Petersen, U. Zeimer, and J. W. Tomm, "How does external feedback cause AlGaAs-based diode lasers to degrade?," *Appl. Phys. Lett.*, vol. 102, no. 2, p. 023502, 2013.

[16] H. Kissel, B. Leonhäuser, J. W. Tomm, M. Hempel, and J. Biesenbach, "Accelerated degradation of high power diode lasers caused by external optical feedback operation," in *2018 IEEE International Semiconductor Laser Conference (ISLC)*, pp. 1–2, Sept 2018.

[17] M. Behringer, *High-Power Diode Laser Technology and Characteristics*, pp. 5–74. New York, NY: Springer New York, 2007.

[18] D. Botez, "Design considerations and analytical approximations for high continuous-wave power, broad-waveguide diode lasers," *Appl. Phys. Lett.*, vol. 74, no. 21, pp. 3102–3104, 1999.

[19] M. Wilkens, H. Wenzel, J. Fricke, A. Maaßdorf, P. Ressel, S. Strohmaier, A. Knigge, G. Erbert, and G. Tränkle, "High-efficiency broad-ridge waveguide lasers," *IEEE Photon. Technol. Lett.*, vol. 30, pp. 545–548, March 2018.

[20] L. Bao, J. Wang, M. Devito, D. Xu, D. Wise, P. Leisher, M. Grimshaw, W. Dong, S. Zhang, K. Price, D. Li, C. Bai, S. Patterson, and R. Martinsen, "Reliability of high performance 9xx-nm single emitter laser diodes," *Proc. SPIE*, vol. 7583, pp. 7583 – 7583 – 10, 2010.

[21] J. Piprek and Z. M. S. Li, "On the importance of non-thermal far-field blooming in broad-area high-power laser diodes," *Appl. Phys. Lett.*, vol. 102, no. 22, p. 221110, 2013.

[22] M. Winterfeldt, *Investigation of slow-axis beam quality degradation in high-power braod area diode lasers.* PhD thesis, Technische Universität Berlin, Fakultät IV, 2018.

[23] M. Winterfeldt, P. Crump, H. Wenzel, G. Erbert, and G. Tränkle, "Experimental investigation of factors limiting slow axis beam quality in 9xx nm high power broad area diode lasers," *J. Appl. Phys.*, vol. 116, no. 6, p. 063103, 2014.

[24] J. N. Walpole, "Semiconductor amplifiers and lasers with tapered gain regions," *Optical and Quantum Electronics*, vol. 28, pp. 623–645, Jun 1996.

[25] C. Zink, A. Maaßdorf, J. Fricke, P. Ressel, M. Maiwald, B. Sumpf, G. Erbert, and G. Tränkle, "Diffraction limited 1064nm monolithic DBR-master oscillator power amplifier with more than 7W output power," in *Proc. SPIE*, vol. 10553, pp. 10553 – 10553 – 7, 2018.

[26] E. S. Kintzer, J. N. Walpole, S. R. Chinn, C. A. Wang, and L. J. Missaggia, "High-power, strained-layer amplifiers and lasers with tapered gain regions," *IEEE Photonics Technology Letters*, vol. 5, pp. 605–608, June 1993.

[27] J. Decker, *Investigation of monolithically integrated spectral stabilization in high-brightness broad area diode lasers.* PhD thesis, Technische Universität Berlin, Fakultät IV, 2018.

[28] P. Crump, C. M. Schultz, H. Wenzel, G. Erbert, and G. Tränkle, "Efficiency-optimized monolithic frequency stabilization of high-power diode lasers," *J. Phys. D: Appl. Phys.*, vol. 46, no. 1, p. 013001, 2013.

[29] J. Decker, P. Crump, J. Fricke, A. Maaßdorf, M. Traub, U. Witte, T. Brand, A. Unger, G. Erbert, and G. Tränkle, "25-W monolithic spectrally stabilized 975-nm minibars for dense spectral beam combining," *IEEE Photon. Technol. Lett.*, vol. 27, pp. 1675–1678, Aug 2015.

[30] M. Haas, A. Killi, C. Tillkorn, S. Ried, M. Ginter, and H. Zimer, "Thin-film filter, wavelength-locked, multi-laser cavity for dense wavelength beam combining of broad-area laser diode bars," *Opt. Lett.*, vol. 40, pp. 3949–3952, Sep 2015.

[31] D. Richter, C. Voigtländer, R. G. Krämer, J. U. Thomas, H. Zimer, A. Tünnermann, and S. Nolte, "Wavelength stabilization of laser diodes by ultrashort pulse written Volume-Bragg-Gratings," in *2015 European Conference on Lasers and Electro-Optics - European Quantum Electronics Conference*, Optical Society of America, 2015.

[32] E. Snitzer, H. Po, F. Hakimi, R. Tumminelli, and B. McCollum, "Double clad, offset core Nd fiber laser," in *Optical Fiber Sensors*, p. PD5, Optical Society of America, 1988.

[33] S. L. Yellen, A. H. Shepard, R. J. Dalby, J. A. Baumann, H. B. Serreze, T. S. Guido, R. Soltz, K. J. Bystrom, C. M. Harding, and R. G. Waters, "Reliability of GaAs-based semiconductor diode lasers: 0.6-1.1 µm," *IEEE J. Quantum Electron.*, vol. 29, pp. 2058–2067, June 1993.

[34] D. Bonsendorf, S. Schneider, J. Meinschien, and J. W. Tomm, "Reliability of high power laser diodes with external optical feedback," *Proc. SPIE*, vol. 9733, pp. 9733 – 9733 – 10, 2016.

[35] N. Chand, W. S. Hobson, J. F. de Jong, P. Parayanthal, and U. K. Chakrabarti, "ZnSe for mirror passivation of high power GaAs based lasers," *Electron. Lett.*, vol. 32, pp. 1595–1596, Aug 1996.

[36] P. Ressel, G. Erbert, U. Zeimer, K. Häusler, G. Beister, B. Sumpf, A. Klehr, and G. Tränkle, "Novel passivation process for the mirror facets of Al-free active-region high-power semiconductor diode lasers," *IEEE Photon. Technol. Lett.*, vol. 17, pp. 962–964, May 2005.

[37] R. W. Lambert, T. Ayling, A. F. Hendry, J. M. Carson, D. A. Barrow, S. McHendry, C. J. Scott, A. McKee, and W. Meredith, "Facet-passivation processes for the improvement of Al-containing semiconductor laser diodes," *Journal of Lightwave Technology*, vol. 24, pp. 956–961, Feb 2006.

[38] I. B. Petrescu-Prahova, P. Modak, E. Goutain, D. Bambrick, D. Silan, J. Riordan, T. Moritz, and J. H. Marsh, "253 mW/µm maximum power density from 9xx nm epitaxial laser structures with d/γ greater than 1 µm," in *2008 IEEE 21st International Semiconductor Laser Conference*, pp. 135–136, Sept 2008.

[39] R. J. Lang, A. G. Larsson, and J. G. Cody, "Lateral modes of broad area semiconductor lasers: theory and experiment," *IEEE J. Quantum Electron.*, vol. 27, pp. 312–320, March 1991.

[40] J. R. Marciante and G. P. Agrawal, "Nonlinear mechanisms of filamentation in broad-area semiconductor lasers," *IEEE J. Quantum Electron.*, vol. 32, pp. 590–596, April 1996.

[41] J. R. Marciante and G. P. Agrawal, "Spatio-temporal characteristics of filamentation in broad-area semiconductor lasers," *IEEE J. Quantum Electron.*, vol. 33, pp. 1174–1179, July 1997.

[42] O. Hess, S. W. Koch, and J. V. Moloney, "Filamentation and beam propagation in broad-area semiconductor lasers," *IEEE J. Quantum Electron.*, vol. 31, pp. 35–43, Jan 1995.

[43] I. Fischer, O. Hess, W. Elsäßer, and E. Göbel, "Complex spatio-temporal dynamics in the near-field of a broad-area semiconductor laser," *EPL (Europhysics Letters)*, vol. 35, no. 8, p. 579, 1996.

[44] J. Martín-Regalado, G. H. M. van Tartwijk, S. Balle, and M. S. Miguel, "Mode control and pattern stabilization in broad-area lasers by optical feedback," *Phys. Rev. A*, vol. 54, pp. 5386–5393, Dec 1996.

[45] S. Wolff, A. Rodionov, V. E. Sherstobitov, and H. Fouckhardt, "Fourier-optical transverse mode selection in external-cavity broad-area lasers: experimental and numerical results," *IEEE J. Quantum Electron.*, vol. 39, pp. 448–458, March 2003.

[46] M. Chi, B. Thestrup, and P. M. Petersen, "Self-injection locking of an extraordinarily wide broad-area diode laser with a 1000-µm-wide emitter," *Opt. Lett.*, vol. 30, pp. 1147–1149, May 2005.

[47] L. Lang, J. J. Lim, S. Sujecki, and E. C. Larkins, "Improvement of the beam quality of a broad-area diode laser using asymmetric feedback from an external cavity," *Opt. Quantum. Electron.*, vol. 40, pp. 1097–1102, Nov 2008.

[48] A. Jechow, V. Raab, R. Menzel, M. Cenkier, S. Stry, and J. Sacher, "1W tunable near diffraction limited light from a broad area laser diode in an external cavity with a line width of 1.7MHz," *Opt. Commun.*, vol. 277, no. 1, pp. 161 – 165, 2007.

[49] L. Borruel, S. Sujecki, P. Moreno, J. Wykes, M. Krakowski, B. Sumpf, P. Sewell, S. C. Auzanneau, H. Wenzel, D. Rodriguez, T. M. Benson, E. C. Larkins, and I. Esquivias, "Quasi-3-D simulation of high-brightness

tapered lasers," *IEEE J. Quantum Electron.*, vol. 40, pp. 463–472, May 2004.

[50] J. J. Lim, S. Sujecki, L. Lang, Z. Zhang, D. Paboeuf, G. Pauliat, G. Lucas-Leclin, P. Georges, R. C. I. MacKenzie, P. Bream, S. Bull, K. Hasler, B. Sumpf, H. Wenzel, G. Erbert, B. Thestrup, P. M. Petersen, N. Michel, M. Krakowski, and E. C. Larkins, "Design and simulation of next-generation high-power, high-brightness laser diodes," *IEEE J. Sel. Topics Quantum Electron.*, vol. 15, pp. 993–1008, May 2009.

[51] H. Wenzel, "Basic aspects of high-power semiconductor laser simulation," *IEEE J. Sel. Topics Quantum Electron.*, vol. 19, pp. 1–13, Sept 2013.

[52] C. Holly, S. Hengesbach, M. Traub, and D. Hoffmann, "Simulation of spectral stabilization of high-power broad-area edge emitting semiconductor lasers," *Opt. Express*, vol. 21, pp. 15553–15567, Jul 2013.

[53] H.-C. Eckstein, *Modenkontrolle in Halbleiterlasern durch monolitisch integrierte mikrooptische Elemente.* PhD thesis, Friedrich-Schiller-Universität Jena, 2013.

[54] J. Piprek, "Self-consistent far-field blooming analysis for high-power fabry-perot laser diodes," in *Proc. SPIE*, vol. 8619, pp. 8619 – 8619 – 8, 2013.

[55] P. Crump, S. Böldicke, C. M. Schultz, H. Ekhteraei, H. Wenzel, and G. Erbert, "Experimental and theoretical analysis of the dominant lateral waveguiding mechanism in 975 nm high power broad area diode lasers," *Semicond. Sci. Technol.*, vol. 27, no. 4, p. 045001, 2012.

[56] C. H. Henry, P. M. Petroff, R. A. Logan, and F. R. Merritt, "Catastrophic damage of $Al_x Ga_{1-x} As$ double-heterostructure laser material," *J. Appl. Phys.*, vol. 50, no. 5, pp. 3721–3732, 1979.

[57] R. Schatz and C. G. Bethea, "Steady state model for facet heating leading to thermal runaway in semiconductor lasers," *J. Appl. Phys.*, vol. 76, no. 4, pp. 2509–2521, 1994.

[58] B. R. Herrero, G. Batko, J. Arias, L. Borruel, I. Esquivias, and R. Gomez-Alcala, "Modeling of facet heating in high-power laser diodes," in *Proc. SPIE*, vol. 3889, pp. 3889 – 3889 – 11, 2000.

[59] D. R. Miftakhutdinov, A. P. Bogatov, and A. E. Drakin, "Catastrophic optical degradation of the output facet of high-power single-transverse-mode diode lasers. 2. Calculation of the spatial temperature distribution and threshold of the catastrophic optical degradation," *Quantum Electron.*, vol. 40, no. 7, p. 589, 2010.

[60] M. Hempel, *Defect mechanisms in diode lasers at high optical output power*. PhD thesis, Humboldt-Universität zu Berlin, Mathematisch-Naturwissenschaftliche Fakultät I, 2013.

[61] J. Souto, J. L. Pura, M. Rodríguez, J. Anaya, A. Torres, and J. Jimenéz, "Mechanisms driving the catastrophic optical damage in high-power laser diodes," in *Proc. SPIE*, vol. 9348, pp. 9348 – 9348 – 7, 2015.

[62] B. Leonhäuser, H. Kissel, J. W. Tomm, M. Hempel, A. Unger, and J. Biesenbach, "High-power diode lasers under external optical feedback," in *Proc. SPIE*, vol. 9348, pp. 9348 – 9348 – 10, 2015.

[63] A. Moser, E. Latta, and D. J. Webb, "Thermodynamics approach to catastrophic optical mirror damage of AlGaAs single quantum well lasers," *Applied Physics Letters*, vol. 55, no. 12, pp. 1152–1154, 1989.

[64] B. Leonhäuser, H. Kissel, A. Unger, B. Köhler, and J. Biesenbach, "Feedback-induced catastrophic optical mirror damage (COMD) on 976nm broad area single emitters with different AR reflectivity," in *Proc. SPIE*, vol. 8965, pp. 896506–896506–10, 2014.

[65] J. Piprek, "Self-consistent analysis of thermal far-field blooming of broad-area laser diodes," *Opt. Quantum. Electron.*, vol. 45, pp. 581–588, Jul 2013.

[66] B. S. Ryvkin and E. A. Avrutin, "Spatial hole burning in high-power edge-emitting lasers: A simple analytical model and the effect on laser performance," *J. Appl. Phys.*, vol. 109, no. 4, pp. 043101–043101–5, 2011.

[67] A. Takeda, R. Shogenji, and J. Ohtsubo, "Spatial-mode analysis in broad-area semiconductor lasers subjected to optical feedback," *Optical Review*, vol. 20, pp. 308–313, Jul 2013.

[68] J. Sigg, "Effects of optical feedback on the light-current characteristics of semiconductor lasers," *IEEE J. Quantum Electron.*, vol. 29, pp. 1262–1270, May 1993.

[69] Z. Chen, L. Bao, J. Bai, M. Grimshaw, R. Martinsen, M. DeVito, J. Haden, and P. Leisher, "Performance limitation and mitigation of longitudinal spatial hole burning in high-power diode lasers," in *Proc. SPIE*, vol. 8277, pp. 8277 – 8277 – 8, 2012.

[70] C. Holly, S. Hengesbach, M. Traub, and D. Hoffmann, "Numerical analysis of external feedback concepts for spectral stabilization of high-power broad-area semiconductor lasers," in *Proc. SPIE*, vol. 8965, pp. 8965 – 8965 – 8, 2014.

[71] A. J. Bennett, R. D. Clayton, and J. M. Xu, "Above-threshold longitudinal profiling of carrier nonpinning and spatial modulation in asymmetric cavity lasers," *J. Appl. Phys.*, vol. 83, no. 7, pp. 3784–3788, 1998.

[72] A. Guermache, V. Voiriot, D. Locatelli, F. Legrand, R. M. Capella, P. Gallion, and J. Jacquet, "Experimental demonstration of spatial hole burning reduction leading to 1480-nm pump lasers output power improvement," *IEEE Photon. Technol. Lett.*, vol. 17, pp. 2023–2025, Oct 2005.

[73] M. Winterfeldt, P. Crump, S. Knigge, A. Maaßdorf, U. Zeimer, and G. Erbert, "High beam quality in broad area lasers via suppression of lateral carrier accumulation," *IEEE Photon. Technol. Lett.*, vol. 27, pp. 1809–1812, Sept 2015.

[74] P. M. Smowton and P. Blood, "The differential efficiency of quantum-well lasers," *IEEE J. Sel. Topics Quantum Electron.*, vol. 3, pp. 491–498, Apr 1997.

[75] S. Rauch, H. Wenzel, M. Radziunas, M. Haas, G. Tränkle, and H. Zimer, "Impact of longitudinal refractive index change on the near-field width of high-power broad-area diode lasers," *Appl. Phys. Lett.*, vol. 110, no. 26, p. 263504, 2017.

[76] M. Peters, V. Rossin, M. Everett, and E. Zucker, "High-power high-efficiency laser diodes at JDSU," in *Proc. SPIE*, vol. 6456, pp. 6456 – 6456 – 11, 2007.

[77] J. Buus, "The effective index method and its application to semiconductor lasers," *IEEE J. Quantum Electron.*, vol. 18, pp. 1083–1089, July 1982.

[78] "BALaser. A software tool for simulation of dynamics in broad area semiconductor lasers." https://www.wias-berlin.de/software/balaser/. Accessed: 2019-05-17.

[79] M. Spreemann, M. Lichtner, M. Radziunas, U. Bandelow, and H. Wenzel, "Measurement and simulation of distributed-feedback tapered master-oscillator power amplifiers," *IEEE J. Quantum Electron.*, vol. 45, pp. 609–616, June 2009.

[80] M. Radziunas and R. Ciegis, "Effective numerical algorithm for simulations of beam stabilization in broad area semiconductor lasers and amplifiers," *Mathematical Modelling and Analysis*, vol. 19, no. 5, pp. 627–646, 2014.

[81] A. Zeghuzi, M. Radziunas, H. Wünsche, J. Koester, H. Wenzel, U. Bandelow, and A. Knigge, "Traveling wave analysis of non-thermal far-field blooming in high-power broad-area lasers," *IEEE J. Quantum Electron.*, vol. 55, pp. 1–7, April 2019.

[82] U. Bandelow, M. Radziunas, J. Sieber, and M. Wolfrum, "Impact of gain dispersion on the spatio-temporal dynamics of multisection lasers," *IEEE J. Quantum Electron.*, vol. 37, pp. 183–188, Feb 2001.

[83] W. Nakwaski, "Static thermal properties of broad-contact double-heterostructure laser diodes," *Opt. Quantum. Electron.*, vol. 15, pp. 513–527, Nov 1983.

[84] L. A. Coldren, S. W. Corzine, and M. L. Masanovic, *A Phenomenological Approach to Diode Lasers*, pp. 45–90. John Wiley & Sons, Inc., 2012.

[85] H. Wenzel, G. Erbert, and P. M. Enders, "Improved theory of the refractive-index change in quantum-well lasers," *IEEE J. Sel. Topics Quantum Electron.*, vol. 5, pp. 637–642, May 1999.

[86] H. Wenzel, P. Crump, A. Pietrzak, X. Wang, G. Erbert, and G. Tränkle, "Theoretical and experimental investigations of the limits to the maximum output power of laser diodes," *New J. Phys.*, vol. 12, no. 8, p. 085007, 2010.

[87] C. Holly, *Modeling of the Lateral Emission Characteristics of High-Power Broad-Area Edge-Emitting Semiconductor Lasers*. PhD thesis, RWTH Aachen University, Aachen, Faculty of Mechanical Engineering, 2019.

[88] A. Demir, M. Peters, R. Duesterberg, V. Rossin, and E. Zucker, "Semiconductor laser power enhancement by control of gain and power profiles," *IEEE Photon. Technol. Lett.*, vol. 27, pp. 2178–2181, Oct 2015.

[89] E. A. Avrutin and B. S. Ryvkin, "Effect of spatial hole burning on output characteristics of high power edge emitting semiconductor lasers: A universal analytical estimate and numerical analysis," *J. Appl. Phys.*, vol. 125, no. 2, p. 023108, 2019.

[90] S. Rauch, M. Haas, and H. Zimer, "Numerical and experimental investigation of near-field narrowing in broad-area laser diodes due to longitudinally asymmetric self-heating," in *2017 IEEE Photonics Conference (IPC)*, pp. 111–112, Oct 2017.

[91] J. Rieprich, M. Winterfeldt, J. Tomm, R. Kernke, and P. Crump, "Assessment of factors regulating the thermal lens profile and lateral brightness in high power diode lasers," in *Proc. SPIE*, vol. 10085, pp. 10085 – 10085 – 10, 2017.

[92] A. Moser, "Thermodynamics of facet damage in cleaved AlGaAs lasers," *Appl. Phys. Lett.*, vol. 59, no. 5, pp. 522–524, 1991.

[93] A. Moser and E. E. Latta, "Arrhenius parameters for the rate process leading to catastrophic damage of AlGaAs-GaAs laser facets," *J. Appl. Phys.*, vol. 71, no. 10, pp. 4848–4853, 1992.

[94] B. S. Ryvkin and E. A. Avrutin, "Effect of carrier loss through waveguide layer recombination on the internal quantum efficiency in large-optical-cavity laser diodes," *J. Appl. Phys.*, vol. 97, no. 11, p. 113106, 2005.

[95] S. Rauch, P. Modak, C. Holly, and H. Zimer, "Beam quality improvement of broad-area laser diodes by symmetric facet reflectivities," in *2018 IEEE International Semiconductor Laser Conference (ISLC)*, pp. 1–2, Sept 2018.

[96] M. Haas, S. Rauch, S. Nagel, L. Irmler, T. Dekorsy, and H. Zimer, "Thin-film filter wavelength-stabilized, grating combined, high-brightness kW-class direct diode laser," *Opt. Express*, vol. 25, pp. 17657–17670, Jul 2017.

[97] A. Demir and M. G. Peters, "Bidirectional long cavity semiconductor laser for improved power and efficiency," May 9 2017. US Patent 9,647,416.

[98] D. Mehuys, R. Lang, M. Mittelstein, J. Salzman, and A. Yariv, "Self-stabilized nonlinear lateral modes of broad area lasers," *IEEE J. Quantum Electron.*, vol. 23, pp. 1909–1920, November 1987.

[99] A. H. Paxton and G. C. Dente, "Filament formation in semiconductor laser gain regions," *J. Appl. Phys.*, vol. 70, no. 6, pp. 2921–2925, 1991.

[100] M. Hempel, J. W. Tomm, M. Baeumler, H. Konstanzer, J. Mukherjee, and T. Elsaesser, "Near-field characteristics of broad area diode lasers during catastrophic optical damage failure," in *Proc. SPIE*, vol. 8432, pp. 8432 – 8432 – 7, 2012.

[101] G. C. Dente, "Low confinement factors for suppressed filaments in semiconductor lasers," *IEEE J. Quantum Electron.*, vol. 37, pp. 1650–1653, Dec 2001.

[102] M. Winterfeldt, P. Crump, S. Knigge, A. Maaßdorf, G. Erbert, and G. Tränkle, "The influence of differential modal gain on the filamentary behavior of broad area diode lasers," in *2015 IEEE Photonics Conference (IPC)*, pp. 567–568, Oct 2015.

[103] M. Chi and P. Michael Petersen, "Spectral properties of a broad-area diode laser with off-axis external-cavity feedback," *Appl. Phys. Lett.*, vol. 103, no. 17, p. 171112, 2013.

[104] B. L. Volodin, S. V. Dolgy, E. D. Melnik, E. Downs, J. Shaw, and V. S. Ban, "Wavelength stabilization and spectrum narrowing of high-power multimode laser diodes and arrays by use of volume Bragg gratings," *Opt. Lett.*, vol. 29, pp. 1891–1893, Aug 2004.

[105] Y. Takiguchi, T. Asatsuma, and S. Hirata, "Effect of the threshold reduction on a catastrophic optical mirror damage in broad-area semiconductor lasers with optical feedback," *Proc. SPIE*, vol. 6104, pp. 6104 – 6104 – 7, 2006.

[106] H. Kissel, P. Wolf, A. Bachmann, C. Lauer, H. König, J. W. Tomm, B. Köhler, U. Strauß, and J. Biesenbach, "Tailored bars at 976 nm for high-brightness fiber-coupled modules," in *Proc. SPIE*, vol. 10086, 2017.

[107] H. C. Casey, D. D. Sell, and K. W. Wecht, "Concentration dependence of the absorption coefficient for n- and p-type GaAs between 1.3 and 1.6 eV," *J. Appl. Phys.*, vol. 46, no. 1, pp. 250–257, 1975.

[108] P. O. Leisher, C. Li, A. K. Jha, K. P. Pipe, J. D. Helmrich, P. Thiagarajan, M. C. Boisselle, S. K. Patra, S. Sezgin, and R. J. Deri, "Feedback-induced failure of high-power diode lasers," *IEEE J. Quantum Electron.*, vol. 54, pp. 1–13, Dec 2018.

[109] C. Li, K. P. Pipe, C. Cao, P. Thiagarajan, R. J. Deri, and P. O. Leisher, "Thermal imaging of high power diode lasers subject to back-irradiance," *Appl. Phys. Lett.*, vol. 112, no. 10, p. 101101, 2018.

[110] S. K. Mandre, I. Fischer, and W. Elsäßer, "Control of the spatiotemporal emission of a broad-area semiconductor laser by spatially filtered feedback," *Opt. Lett.*, vol. 28, pp. 1135–1137, Jul 2003.

[111] T. Tachikawa, R. Shogenji, and J. Ohtsubo, "Observation of multi-path interference in broad-area semiconductor lasers with optical feedback," *Optical Review*, vol. 16, pp. 533–539, Sep 2009.

[112] S. Rauch, C. Holly, and H. Zimer, "Catastrophic optical damage in 950-nm broad-area laser diodes due to misaligned optical feedback and injection," *IEEE J. Quantum Electron.*, vol. 54, pp. 1–7, Aug 2018.

[113] Y. Mitsuhashi, J. Shimada, and S. Mitsutsuka, "Voltage change across the self-coupled semiconductor laser," *IEEE J. Quantum Electron.*, vol. 17, pp. 1216–1225, Jul 1981.

[114] P. Epperlein, *Semicondutor Laser Engineering, Reliability and Diagnostics: A Practical Approach to High Power and Single Mode Devices*. Jon Wiley & Sons Ltd, 2013.

[115] D. Pierścińska, "Thermoreflectance spectroscopy—analysis of thermal processes in semiconductor lasers," *J. Phys. D: Appl. Phys.*, vol. 51, p. 013001, nov 2017.

[116] P.-W. Epperlein, "Micro-temperature measurements on semiconductor laser mirrors by reflectance modulation: A newly developed technique for laser characterization," *Japanese Journal of Applied Physics*, vol. 32, no. 12R, p. 5514, 1993.

[117] H. Yao, P. G. Snyder, and J. A. Woollam, "Temperature dependence of optical properties of GaAs," *J. Appl. Phys.*, vol. 70, no. 6, pp. 3261–3267, 1991.

[118] K. Pierściński, D. Pierścińska, and M. Bugajski, "Quantification of thermoreflectance temperature measurements in high-power semiconductor devices - lasers and laser bars," *Microelectron. J.*, vol. 40, no. 9, pp. 1373 – 1378, 2009. Quality in Electronic Design 2nd IEEE International Workshop on Advances in Sensors and Interfaces Thermal Investigations of ICs and Systems.

[119] S. Dilhaire, S. Grauby, and W. Claeys, "Thermoreflectance calibration procedure on a laser diode: application to catastrophic optical facet damage analysis," *IEEE Electron Device Lett.*, vol. 26, pp. 461–463, July 2005.

[120] M. Suzuki, T. Yamamoto, Y. Katayama, S. Kuwata, and T. Tanaka, "Light absorption by metals with porous surface layer formed by oxidization-reduction treatment," *Mater. Trans.*, vol. 53, no. 9, pp. 1556–1562, 2012.

[121] A. K. Jha, C. Li, K. P. Pipe, M. T. Crowley, D. B. Fullager, J. D. Helmrich, P. Thiagarajan, R. J. Deri, E. Feigenbaum, R. B. Swertfeger, and P. O. Leisher, "Thermoreflectance imaging of back-irradiance heating in high power diode lasers at several operating wavelengths," *IEEE J. Sel. Topics Quantum Electron.*, vol. 25, pp. 1–13, Nov 2019.

[122] M. Hempel, J. W. Tomm, F. L. Mattina, I. Ratschinski, M. Schade, I. Shorubalko, M. Stiefel, H. S. Leipner, F. M. Kießling, and T. Elsaesser, "Microscopic origins of catastrophic optical damage in diode lasers," *IEEE J. Sel. Topics Quantum Electron.*, vol. 19, p. 1500508, July 2013.

[123] M. Hempel, F. L. Mattina, J. W. Tomm, U. Zeimer, R. Broennimann, and T. Elsaesser, "Defect evolution during catastrophic optical damage of diode lasers," *Semicond. Sci. Technol.*, vol. 26, p. 075020, Apr 2011.

[124] U. Menzel, "Self-consistent calculation of facet heating in asymmetrically coated edge emitting diode lasers," *Semicond. Sci. Technol.*, vol. 13, no. 3, p. 265, 1998.

Acknowledgments

In the following I would like to appreciate the support of people who accompanied me during the time of my doctorate studies.

First of all, I would like to thank Prof. Dr. Günther Tränkle for giving me the great opportunity to prepare this dissertation in the collaboration of the Ferdinand-Braun-Institut, Leibniz-Institut für Höchstfrequenztechnik (FBH) and TRUMPF. His very competent and valuable feedback helped immensely to structure my work packages. Furthermore, I appreciate Prof. Dr. Thomas Dekorsy and Prof. Dr. Michael Kneissl reviewing this dissertation.

I am deeply indebted to Dr. Hagen Zimer for his comprehensive, inspiring and competent supervision: His clear and vivid understanding of physics has become an omnipresent benchmark and shaped the argumentation of this work. Besides his professional assistance and his great ingenuity, I am grateful for his supporting words of advice exchanged during several long-distance runs.

I would like to thank my colleague Carlo Holly for his very competent and profound support in the field of numerical simulations. I have fond memories of the trips with him to New York city and Princeton bristled with long discussions on broad-area lasers.

My thanks go also to Dr. Hans Wenzel from FBH and Dr. Mindaugas Radziunas from the Weierstrass Institute for Applied Analysis and Stochastics for sharing their excellent experience of broad-area laser modeling and simulation.

With respect to the experimental side of this work, I want to express my sincere gratitude to the team of TRUMPF Photonics Inc., Cranbury NJ for providing me excellent laser devices. Special thanks go to Dr. Prasanta Modak, Ayesha Jamil and Dr. Stewart McDougall for their technical advice. Moreover, I appreciate the great financial and scientific freedom provided by Steffen Ried and Dr. Alexander Killi from TRUMPF Laser GmbH, Schramberg (TLS).

Furthermore, I thank Dr. Ute Zeimer from FBH for providing the cathodoluminescence scans. My tanks are extended to the laser application group of Elke Kaiser for the preparation of submounts.

I am glad to have had a very welcoming and supportive working environment provided by my co-workers at TLS. I am thankful for the motivational atmosphere in the laboratory created by Simon Nagel, Markus Ginter and Matthias Haas and the help of these colleagues proof-reading my manuscripts. I would further like to thank my fellow PhD students at FBH for their positive feedback and welcoming tours through Berlin.

Moreover, I am pleased with people that have introduced me to the life in the Black Forest area. The bike tours with and by Markus Ginter through the high-mountain valleys as well as the barbecues with Simon Nagel have always been a refreshing change from everyday life.

Last but not least, I am most deeply indebted to my beloved girlfriend Karoline Preis for her love, patience and tremendous support over the past years. Especially, her infectious warm-heartedness has become a reliant adviser within my life. I am exceptionally grateful for every moment we spent and will spend together. I would like to extend my gratitude further to my and Karoline's parents and family for their steady support and confirmation.

Innovationen mit Mikrowellen und Licht
Forschungsberichte aus dem Ferdinand-Braun-Institut, Leibniz-Institut für Höchstfrequenztechnik

Herausgeber: Prof. Dr. G. Tränkle, Prof. Dr.-Ing. W. Heinrich

Band 1:
Thorsten Tischler
Die Perfectly-Matched-Layer-Randbedingung in der
Finite-Differenzen-Methode im Frequenzbereich:
Implementierung und Einsatzbereiche
ISBN: 3-86537-113-2, 19,00 EUR, 144 Seiten

Band 2:
Friedrich Lenk
Monolithische GaAs FET- und HBT-Oszillatoren
mit verbesserter Transistormodellierung
ISBN: 3-86537-107-8, 19,00 EUR, 140 Seiten

Band 3:
R. Doerner, M. Rudolph (eds.)
Selected Topics on Microwave Measurements,
Noise in Devices and Circuits, and Transistor Modeling
ISBN: 3-86537-328-3, 19,00 EUR, 130 Seiten

Band 4:
Matthias Schott
Methoden zur Phasenrauschverbesserung von
monolithischen Millimeterwellen-Oszillatoren
ISBN: 978-3-86727-774-0, 19,00 EUR, 134 Seiten

Band 5:
Katrin Paschke
Hochleistungsdiodenlaser hoher spektraler Strahldichte
mit geneigtem Bragg-Gitter als Modenfilter (α-DFB-Laser)
ISBN: 978-3-86727-775-7, 19,00 EUR, 128 Seiten

Band 6:
Andre Maaßdorf
Entwicklung von GaAs-basierten Heterostruktur-Bipolartransistoren
(HBTs) für Mikrowellenleistungszellen
ISBN: 978-3-86727-743-3, 23,00 EUR, 154 Seiten

Band 7:
Prodyut Kumar Talukder
Finite-Difference-Frequency-Domain Simulation of Electrically
Large Microwave Structures using PML and Internal Ports
ISBN: 978-3-86955-067-1, 19,00 EUR, 138 Seiten

Band 8:
Ibrahim Khalil
Intermodulation Distortion in GaN HEMT
ISBN: 978-3-86955-188-3, 23,00 EUR, 158 Seiten

Band 9:
Martin Maiwald
Halbleiterlaser basierte Mikrosystemlichtquellen für die Raman-Spektroskopie
ISBN: 978-3-86955-184-5, 19,00 EUR, 134 Seiten

Band 10:
Jens Flucke
Mikrowellen-Schaltverstärker in GaN- und GaAs-Technologie
Designgrundlagen und Komponenten
ISBN: 978-3-86955-304-7, 21,00 EUR, 122 Seiten

Cuvillier Verlag
Internationaler wissenschaftlicher Fachverlag

Innovationen mit Mikrowellen und Licht
Forschungsberichte aus dem Ferdinand-Braun-Institut,
Leibniz-Institut für Höchstfrequenztechnik

Herausgeber: Prof. Dr. G. Tränkle, Prof. Dr.-Ing. W. Heinrich

Band 11:   **Harald Klockenhoff**
Optimiertes Design von Mikrowellen-Leistungstransistoren
und Verstärkern im X-Band
ISBN: 978-3-86955-391-7, 26,75 EUR, 130 Seiten

Band 12:   **Reza Pazirandeh**
Monolithische GaAs FET- und HBT-Oszillatoren
mit verbesserter Transistormodellierung
ISBN: 978-3-86955-107-8, 19,00 EUR, 140 Seiten

Band 13:   **Tomas Krämer**
High-Speed InP Heterojunction Bipolar Transistors
and Integrated Circuits in Transferred Substrate Technology
ISBN: 978-3-86955-393-1, 21,70 EUR, 140 Seiten

Band 14:   **Phuong Thanh Nguyen**
Investigation of spectral characteristics of solitary
diode lasers with integrated grating resonator
ISBN: 978-3-86955-651-2, 24,00 EUR, 156 Seiten

Band 15:   **Sina Riecke**
Flexible Generation of Picosecond Laser Pulses in the Infrared and
Green Spectral Range by Gain-Switching of Semiconductor Lasers
ISBN: 978-3-86955-652-9, 22,60 EUR, 136 Seiten

Band 16:   **Christian Hennig**
Hydrid-Gasphasenepitaxie von versetzungsarmen und freistehenden
GaN-Schichten
ISBN: 978-3-86955-822-6, 27,00 EUR, 162 Seiten

Band 17:   **Tim Wernicke**
Wachstum von nicht- und semipolaren InAlGaN-Heterostrukturen
für hocheffiziente Licht-Emitter
ISBN: 978-3-86955-881-3, 23,40 EUR, 138 Seiten

Band 18:   **Andreas Wentzel**
Klasse-S Mikrowellen-Leistungsverstärker mit GaN-Transistoren
ISBN: 978-3-86955-897-4, 29,65 EUR, 172 Seiten

Band 19:   **Veit Hoffmann**
MOVPE growth and characterization of (In,Ga)N quantum structures
for laser diodes emitting at 440 nm
ISBN: 978-3-86955-989-6, 18,00 EUR, 118 Seiten

Band 20:   **Ahmad Ibrahim Bawamia**
Improvement of the beam quality of high-power broad area
semiconductor diode lasers by means of an external resonator
ISBN: 978-3-95404-065-0, 21,00 EUR, 126 Seiten

Cuvillier Verlag
Internationaler wissenschaftlicher Fachverlag

Innovationen mit Mikrowellen und Licht
Forschungsberichte aus dem Ferdinand-Braun-Institut,
Leibniz-Institut für Höchstfrequenztechnik

Herausgeber: Prof. Dr. G. Tränkle, Prof. Dr.-Ing. W. Heinrich

Cuvillier Verlag
Internationaler wissenschaftlicher Fachverlag

Innovationen mit Mikrowellen und Licht
Forschungsberichte aus dem Ferdinand-Braun-Institut,
Leibniz-Institut für Höchstfrequenztechnik

Herausgeber: Prof. Dr. G. Tränkle, Prof. Dr.-Ing. W. Heinrich

Band 41: **Thi Nghiem Vu**
Development and analysis of diode laser ns-MOPA systems
for high peak power application
ISBN: 978-3-7369-9480-5, 38,80 EUR, 138 Seiten

Band 42: **Christian Bansleben**
Differentieller Mikrowellen-Leistungsoszillator für die Realisierung
ultrakompakter Plasmaquellen in Matrixanordnung
ISBN: 978-3-7369-9530-7, 34,90 EUR, 136 Seiten

Band 43: **Martin Winterfeldt**
Investigation of slow-axis beam quality degradation
in high-power broad area diode lasers
ISBN: 978-3-7369-9733-2, 39,90 EUR, 158 Seiten

Band 44: **Jonathan Decker**
Investigation of monolithically integrated spectral stabilization
in high-brightness broad area diode lasers
ISBN: 978-3-7369-9798-1, 49,50 EUR, 174 Seiten

Band 45: **Andreea Cristina Andrei**
Untersuchung und Optimierung robuster und hochlinearer
rauscharmer Verstärker in GaN-Technologie
ISBN: 978-3-7369-9810-0, 39,90 EUR, 148 Seiten

Band 46: **Peng Luo**
GaN HEMT Modeling Including Trapping Effects Based on
Chalmers Model and Pulsed S-Parameter Measurements
ISBN: 978-3-7369-9906-0, 48,00 EUR, 160 Seiten

Band 47: **Jörg Jeschke**
Entwicklung von optisch pumpbaren UVC-Lasern auf AlGaN-Basis
Chalmers Model and Pulsed S-Parameter Measurements
ISBN: 978-3-7369-9918-3, 44,90 EUR, 176 Seiten

Band 48: **Nikolai Wolff**
Wideband GaN Microwave Power Amplifiers with Class-G
Supply Modulation
ISBN: 978-3-7369-9931-2, 44,90 EUR, 170 Seiten

Band 49: **Carlo Frevert**
Optimization of broad-area GaAs diode lasers for high powers
and high efficiences in the temperature range 200-220 K
ISBN: 978-3-7369-9944-2, 44,90 EUR, 174 Seiten

Band 50: **Bassem Arar**
GaAs-based components for photonic integrated circuits
ISBN: 978-3-7369-9976-3, 43,60 EUR, 152 Seiten

Cuvillier Verlag
Internationaler wissenschaftlicher Fachverlag

Innovationen mit Mikrowellen und Licht
Forschungsberichte aus dem Ferdinand-Braun-Institut,
Leibniz-Institut für Höchstfrequenztechnik

Herausgeber: Prof. Dr. G. Tränkle, Prof. Dr.-Ing. W. Heinrich

Band 51: **Mahmoud Tawfieq**
Development and characterisation of a diode laser based tunable
high-power MOPA system
ISBN: 978-3-7369-9983-1, 44,90 EUR, 170 Seiten

Band 52: **Sebastian Preis**
Hocheffiziente frequenzagile Mikrowellen-Leistungsverstärker
auf Basis von Verbindungshalbleitern und Ferroelektrika
ISBN: 978-3-7369-7004-5, 54,00 EUR, 136 Seiten

Band 53: **Simon Fleischmann**
Materialaspekte der Hydridgasphasenepitaxie
von Aluminiumgalliumnitrid
ISBN: 978-3-7369-7029-8, 41,80 EUR, 160 Seiten

Band 54: **Erhan Ersoy**
Optimierung von koplanaren GaN-MMIC-Leistungsverstärkern
im X-Band
ISBN: 978-3-7369-7157-8, 39,90 EUR, 154 Seiten